SpringerBriefs in Applied Sciences and Technology

Series editor

Andreas Öchsner, Southport, Australia

For further volumes:
http://www.springer.com/series/8884

Marc Méquignon · Hassan Ait Haddou

Lifetime Environmental Impact of Buildings

Marc Méquignon
UPS- IUT-LERASS ERPURS
Toulouse
France

Hassan Ait Haddou
Laboratoire Innovation Formes
   Architectures Milieux
Ecole Nationale Supérieure d'Architecture
   de Montpellier
Montpellier
France

ISSN 2191-530X   ISSN 2191-5318   (electronic)
ISBN 978-3-319-06640-0   ISBN 978-3-319-06641-7   (eBook)
DOI 10.1007/978-3-319-06641-7
Springer Cham Heidelberg New York Dordrecht London

Library of Congress Control Number: 2014938216

Printed on acid-free paper

Springer is part of Springer Science+Business Media (www.springer.com)

Acknowledgments

The first author would like to think Carole, Florian, and Marine Méquignon for their permanent support during the period of developing this book. He would like to thanks the LERASS-laboratory members.

The second author would like to sincerely thank his wife Leila and his son Aymen for here energetic and kind support during the book development and editorial process.

Special thanks to Springer's Global Editorial Board, Anthony Doyle, Ashok Arumairaj, Gabriella Anderson for their support and help during the editorial process.

Many thanks to friends and colleagues, Marc, Vinicius, Nicolas, Philippe, Laurent, and to the members of the LIFAM laboratory.

Contents

Part I   Background

1   The Construction Industry and Lifespan 3
   1.1   Ageing of a Building 3
         1.1.1   Ageing 3
         1.1.2   Obsolescence 4
   1.2   Buildings, Construction Products and Lifespan.................. 5
         1.2.1   Technical Lifespan and Tools for Evaluating It 5
         1.2.2   Functional Lifespans of Products and Buildings 9
   1.3   Factors Influencing Lifespan 12
         1.3.1   Impact of Maintenance on Lifespan................... 12
         1.3.2   Maintenance and Maintenance Rules................... 12
         1.3.3   Adaptability 13
         1.3.4   Product Reuse and Recycling. 13
   1.4   Tools for Assessing Performance 16
         1.4.1   Taking Account of Lifespan Using Standards 16
         1.4.2   Lifespan Consideration in Tools of Impact Assessment. 17
   1.5   General Analysis. 18
         1.5.1   The Window Example 18
         1.5.2   Recycling and Recovery........................... 19
         1.5.3   EUROCODE 0 19
         1.5.4   Concerning Maintenance 20
         1.5.5   Adaptability 20
         1.5.6   Optimization...................................... 20
         1.5.7   Need 21
   1.6   Conclusion 21
   References.. 22

2   Building and Sustainable Development 25
   2.1   Background 25
         2.1.1   Definitions 26
         2.1.2   Principle: Analysis of Life Cycle 27
         2.1.3   Eco-Design and Eco-Building Materials 28

2.2 Indicators and Sustainable Development Data.................. 30
    2.2.1 Indicators and Indices............................... 30
    2.2.2 Emissions from the Sector and Its Products............. 33
    2.2.3 The Characteristics of the Data and Data Sources........ 36
2.3 Tools for Ecological Performance 39
    2.3.1 Standards and Guides............................... 39
    2.3.2 The Analytical Tools for Evaluating the Performance...... 41
    2.3.3 The Models..................................... 42
References... 42

3 Research Analysis 45
  3.1 Conclusion 47
  References... 47

Part II Method and Application

4 Method.. 51
  4.1 Wall Unit 51
      4.1.1 Delimitation of the System........................... 51
      4.1.2 Scales of Evaluation................................ 54
      4.1.3 Summary 54
  4.2 Choice of Indicators and Means Employed 55
      4.2.1 General Principles of Evaluation 55
      4.2.2 Choice of Indicators 56
      4.2.3 Availability and Choice of Data Sources 56
      4.2.4 Data Sources..................................... 57
      4.2.5 Contradictory Data in Environmental Databases 59
  4.3 Presentation of the Method................................. 63
      4.3.1 Analysis of Performance of the Wall Unit and the House ... 63
      4.3.2 Taking Impact of Utilization into Account............... 64
      4.3.3 Development of the Method........................... 65
  4.4 Conclusion 69
  References... 70

5 Application... 71
  5.1 Data Selected 71
  5.2 Indicator Values 71
  5.3 Definition of the Element Under Study 78
      5.3.1 Chosen Solutions (Phase 2) 80
  5.4 Environmental Data: Values of GHG Indicator 82

6 Results and Analyses.. 83
  6.1 Cumulative GHG Emissions for the Walls (Phase 5)............ 83
  6.2 Relative Proportions of Emissions from Insulation
      and Coatings for the Duration of the Function................. 86

    6.2.1 Solutions Having the Best Performance................ 86
    6.2.2 Second-Order Solutions............................. 87
  6.3 Demonstration of the Uncertainties Inherent
    in the Official Indicators................................. 88
  6.4 Impact of Service Life of Technical Solutions
    on GHG Emission....................................... 89
    6.4.1 Evolution of Emissions (Phase 6).................... 89
  6.5 Comparison of Technical Solutions......................... 91
  6.6 Conclusion.. 92

Part III Conditions for Generalization and Prospects

7 Conditions for Generalizing the Approach...................... 95
  7.1 Analysis.. 95
  7.2 Modelling Elements...................................... 95
    7.2.1 Evolution of the Impact of a Product i Over Time........ 95
    7.2.2 Evolution of the Impact of a Product i According
      to Its Service Life.............................. 96
    7.2.3 Analyses.. 96
    7.2.4 Impact of Service Life and Technological
      Improvements.................................... 97
    7.2.5 Comparison of Products........................... 97

8 Limits.. 99

9 Interest and Prospects...................................... 101
  9.1 Contributions to a Sustainable Development Approach.......... 101
  9.2 Rescarch Prospects...................................... 102
    9.2.1 Impact of Lifetime for a Building.................... 102
    9.2.2 Multidisciplinary Reflection for a Sustainable
      Development Approach............................ 103
    9.2.3 Indicators and Evaluation Tools...................... 103
    9.2.4 Deeper Understanding of the Relative Contributions
      of the Component Parts of a Building................. 103
    9.2.5 Recycling....................................... 103
    9.2.6 Adaptability..................................... 104
    9.2.7 Total Number of Buildings and Demand................ 104
    9.2.8 Impact at Different Scales......................... 105
  References.. 105

Conclusion.. 107

Appendix 1: Référentiels... 109

Appendix 2: Tools of Sustainable Development Assessment............ 117

**Appendix 3: Management Protocol Database on
            Environmental and Health Declaration of
            Construction Products (INIES)** 121

Abstract

This book aims to highlight the impact of the lifespan of the buildings on greenhouse gas emissions. The authors conduct a thorough study of the state of the art which defines the lifetime of buildings and the issue of sustainable development when applied to this area. To fulfill the objective, the authors focus on the assessment of greenhouse gas emissions produced by the walls of buildings according to their lifespan. These assessments take account of the construction, maintenance, and end of life. The contribution of the utilization phase must be equivalent for all technical solutions for a given usage function. The methodology is described by considering a unit area of wall (i.e., 1 square meter), determining a long service life, choosing technical solutions in agreement with the specifications, establishing the lifespan of each technical solution according to experts, finding the corresponding greenhouse gas index from an appropriate database, and finally modeling the evolution of these indicators with time.

Several technical solutions (concrete, brick, stone, wood, aerated concrete) are considered and lifespans range from a few years to centuries.

The results of this analysis suggest and quantify the important impact of lifespan on greenhouse gas emission indicators. For example, the best technical solution for a short lifetime can be the worst on a longer duration and vice versa. Therefore, since the lifespan of a product is very difficult to determine objectively, it is considered as a variable. The numerical results presented point out the need to revisit the current life cycle tools.

The conditions and limits of the generalization of the method and results are presented and analyzed.

The authors hope to be, through this book, attracted interest of the scientific community for a new impact on the sustainability performance factor, namely, lifespan.

Introduction

The purpose of this book is to initiate reflection on the question of the methods that can be used to evaluate the impact of the lifespan of buildings on sustainable development.

The impacts of a building's phase of use have been the subject of many investigations in the last few years and progress for this phase, at least with regard to new constructions. Therefore, it is time to evaluate the question of performance in terms of sustainable development, taking the lifespan factor into consideration. To attain this very ambitious objective, it will be indispensable to break the problems down and reduce them, for two essential reasons. On the one hand, a building is a very complex system and the tools currently available are unable to take a selected lifespan factor into consideration when measuring the building's performance. On the other, the very concept of sustainable development is still very vague as regards the point at issue and, having no integrated indicators, requires a reduced number of evaluation criteria.

For these reasons, we will restrict ourselves to the study of one component of the building evaluated by means of a selected indicator. The principal limits of the proposed method are discussed, as are the conditions for its generalization to the entire building.

The book is divided into three parts. The first presents the background discussing the relationship between a building and (1) its lifespan and (2) sustainable development, concluding with a comprehensive analysis. The second part presents the proposed method and its application with some results and observations.

Finally, the third part outlines the conditions under which the method can be generalized to the whole building, discusses its limits and indicates prospects for further research.

Part I
Background

Chapter 1
The Construction Industry and Lifespan

Abstract This chapter presents the state of the art of the lifespan of buildings and their components. It describes the mechanisms of ageing, lifespan assessment tools and its consequences. The actual life of a building is generally the shortest amongst the technical or functional life of the building. The technical life is linked to product quality and functional life depends on the owner and user care and practices. Maintenance is a common factor that may alter both lifespans. Finally, other exogenous factors could be the cause for demolition.

Keywords Building • Tools • Lifespan

1.1 Ageing of a Building

We use the word "building" here as "construction which is used as shelter for men, animals, and objects" (Choay and Merlin 2010). When it concerns the entire activity of the building, the term used is "building domain".

1.1.1 Ageing

The ageing of an object is defined as its increased risk of mortality over time (Miller 2002). Objects can "live" to be much older than living species. In fact, engineering jargon in general draws heavily on that of the living world: ageing of course, but also pathology, level of disorder, loss of stamina, reduced lifespan, shape memory, fatigue, etc.

André Zaoui, in letter No. 3 2002/printemps of the Academy of Sciences (Zaoui 2002), explains that manufactured products "are subjected to constraints that vary between the time of manufacture and the time of use". The constituent materials are under the influence of "an implacable driving force generated by

M. Méquignon and H. Ait Haddou, *Lifetime Environmental Impact of Buildings*,
SpringerBriefs in Applied Sciences and Technology, DOI: 10.1007/978-3-319-06641-7_1,
© The Author(s) 2014

thermodynamic imbalance, which will inevitably act to reduce the imbalance by changing the inner architecture of matter and the properties for which it is responsible: ageing"... The process is called "thermal ageing" because in the background, though apparently in charge, lies omnipotent temperature while what is really accountable is how thermal activation is regulated.

Ageing is also the reaction of a product to constant use, which can correspond to an increase in its internal temperature. This is the phenomenon of wear. Even though certain materials, such as concretes, have recently been designed to be repaired (Wang et al. 2006), the fact that their lifespan may be extended does not prevent them from being subject to ageing.

Concerning the ageing of buildings and blocks, the mechanism of degradation, linked to the joiner's status and associated with its maintenance, has been very accurately described (Levy and Saint Raymond 1992). This ageing is largely associated with the social evolution of the occupants: the former occupant is inevitably replaced by an occupant accepting a slightly more degraded state.

This in turn causes the neighbour to adopt a similar process and the building suffers gradual degradation as it receives only minimal maintenance.

In addition, demographic analyses can also lead to a description of the "mechanical" phenomenon of ageing of a city and buildings. People arriving from new districts have a relatively homogeneous social profile which gives rise to a similar ageing of its population and consequently of its social life.

1.1.2 Obsolescence

In the field of consumer goods, there is a phenomenon of obsolescence. In other words, the product, independently of its physical characteristics, no longer satisfies evolving needs. Under cover of quality measures, using tools such as value analysis, the lifespan of the objects that surround us is adjusted to the strict satisfaction of consumer needs. As a result of more effective marketing tools, the perceived needs change ever faster and the commercial life cycles of products are becoming shorter in all areas. For example, in Europe, in the automobile industry, the life cycle of a model lasted 14 years in the 1960s but had been reduced to 10 years in 1975 (Dupuy and Gerin 1975) and is about 6 years today, including an intermediate restyling. The upshot of this is that the consumer renews his purchase after a shorter delay. This leads certain critical analysts to talk about planned obsolescence, a concept that consists of programming the end of the lifetime of a product during its design or manufacture. In this way, the sector can ensure that the user will be obliged to buy a new one. On the other hand, marketing experts associate this phenomenon of obsolescence with the "creation of new needs". This position is validated by the scrapping of domestic products before their technical failure (Cooper 2004). In addition, the defenders of the industrial activity in question will retort that this established fact is the corollary of optimized costs and thus of reduced selling prices.

In the building sector, thanks to knowledge becoming increasingly refined and clear, buildings often span centuries. A walk through our old historical centres is sufficient to show us that a large proportion of the buildings live on, at least as regards their structure. This observation raises questions on this phenomenon of obsolescence, perfectly described by Lemer (1996). The characteristics of a great number of buildings, associated with the decisions of their successive owners, have enabled them to adapt to the times, either continuously or in spurts, thus pushing back the consequences of possible obsolescence.

1.2 Buildings, Construction Products and Lifespan

Research shows that household goods are rarely discarded or destroyed due to technical failure or for fear of it (Cooper 2004). Although buildings are probably not everyday consumer goods, it seems necessary to study the reasons that can lead to the end of their lifetimes. The object of this section is to list, as completely as possible, the factors that may modify the course of a building's life cycle and to analyse their mechanisms. To do this, it is necessary to dissociate the "technical" lifespan, and the "functional" lifespan.

1.2.1 Technical Lifespan and Tools for Evaluating It

The technical lifespan results from wear related to use and the effect of time. The question of the lifespan of structures has been under discussion since 1970. The question of evaluating the lifespan of buildings, and especially of the products that compose them, was first tackled in the 1990s. Based on the structural and functional analysis of the components of the building, the causes of failure are established and the lifespans evaluated. As far as the tools for evaluating lifespan are concerned, great progress has been made. The methods and tools for evaluating lifespans using statistical calculations and failure models have been minutely described and are precisely implemented (Lair et al. 2003; Talon 2006b).

1.2.1.1 Technical Lifespan of Products and Buildings

The technical lifespan of the products composing a building can be evaluated with a specific probability. Talon (2006b) established that the technical lifespan of a double-glazed PVC window ranges between 23 and 27 years. Failure of the elements of a building do not imply the end of the lifetime of the building itself. A systemic analysis of the building makes it possible to produce an exhaustive list of the elements whose failure ends the lifetime of the building. Table 1.1 detailed hereafter allows such an analysis.

In summary, concerning the impact of the lifespan of the building's components, it is only when the reliability of the foundations, the load-bearing walls or the floors is doubtful that the building is at risk of destruction. The probability of demolition is high only when damage is generalized. When the damage is partial,

Table 1.1 Analysis of the technical risks concerning building elements

Component	Function	Risks	Link with other components	Risk of demolition
Foundations	Load distribution on the support	Mechanical failure	Not exchangeable	X
Structure	Transmit higher loads, and its own load, to the ground	Failure	Not exchangeable	X
GW insulation	Reduce exchanges between inside and outside	Deterioration and loss of efficiency	Exchangeable	
Exterior coating	Protect from weather damage and outside eyes	Increased? permeability	Exchangeable	
		Degraded appearance	Exchangeable	
Plaster coating	Support the finish Protect the insulation	Damage	Exchangeable	
Floors	Allow the use functions of daily life	Mechanical failure	Not exchangeable	X
		Decreased? standard of comfort		
Partition walls	Internal distribution of rooms	Changes in needs	Exchangeable	
		Mechanical failure		
Framework	Support coverage	Mechanical failure	Exchangeable	
Roof covering	Protect the entire house	Mechanical failure	Exchangeable	
		Destruction by wind uplift		
Outside frames	Promote the illumination Allow ventilation Insulation from outside (thermal and phonic)	Sealing reduction Mechanical failure	Exchangeable	
Finish	Improve appearance	Dilapidated appearance	Exchangeable	
Equipment	Improve comfort	Failure	Exchangeable	
Accidental internal risk on elements				
Fire				X
Explosions				X

it may still be possible for the building to be repaired. The other components, the lifespans of which are sometimes very short, do not have any direct impact on the durability of the building. For instance, door and window frames will be changed as soon as they no longer ensure that the building is waterproof or that it prevents intrusions. The roof will be checked, even removed and replaced … Finally, there are two internal technical risks that can lead to the end of its lifetime: the risks of fire or explosion. Thus it is the mechanical lifespan of the structure that determines the technical lifespan of the building. For calculation of the structural dimensions,

EUROCODE 0, which came into force on March 1st, 2010, defines the reference lifespan of buildings according to type of use. For the most common buildings, duration is fixed at 50 years.

1.2.1.2 Evaluation Tools

Several tools concerning the lifespan have been produced in recent decades.

The studies carried out have given rise to a number of standards and regulations for lifespan calculation including but not limited to:

- 1978: U.S. standard ASTM E 632-82 Subcommittee E06.22 of the American Society for Testing and Materials.
- 1988: French standard NF X 60-500
- 1992: British Standard BS 7543
- 1993: English version of the guide from the Architectural Institute of Japan
- 1995: Guide CSA S478-95 from the Canadian Standards Association
- 2000: ISO 15686—Buildings and constructed assets—Service life schedule
- 2010: Eurocode 0, specifying the methods for calculating the structure of a building and its lifespan.

The development of tools made it necessary to draw up the appropriate definitions, which are summarized in the standard ISO 15686-1—Buildings and constructed assets—Design taking into account the lifespan—Part 1: General principles and scope. These definitions are as follows:

Lifespan of design: lifespan sought by the designer, for example the design lifespan of the Millau viaduct is 120 years.

Lifespan of reference: expected lifespan of a product, a component, an assembly or a system for a particular construction. This lifetime may be obtained on the basis of accelerated ageing tests.

Estimated lifespan: "expected lifetime of a building or building components for a set of specific conditions of service determined from reference lifetime data taking into account differences in conditions of service".

- Residual lifespan: "remaining lifespan from the time in question".

In summary, the ISO 15686 standard provides a collectively accepted definition. Lifespan is defined as "the period after implementation for which a building or building component meets or exceeds the performance requirements."

Finally, different methods have been developed for assessing technical lifespan. These methods are presented in Table 1.2 (Talon 2006a).

These approaches are distinguished from one another by the type of elements studied, the type and number of experimental data, the type of approach (random, semi deterministic, deterministic), the processing times and the precision of the results.

Talon voiced two criticisms of these approaches: they evaluate only one element at a time, and the lifespans computed are subject to errors, inaccuracies and incompleteness, stemming mainly from the acquisition modes and uncertainties.

Table 1.2 Comparison of methods for computation of lifespan

Approach	Data used	Scales	Benefits	Limits
Experimental approach	Experimental data according to field exposures in experimental buildings, short term accelerated	Material Product Building	Possible to take several random parameters into account	Needs probabilistic data modeling
Statistical approach	All types of experimental data	Material Product Building	Possible to take several random parameters into account	Needs a large amount of experimental data
Models of ageing mechanism	Experimental data from field exposure in experimental buildings, short term accelerated	Material	Theoretical modelling of material behaviour	The model is more accurate unless it can be extrapolated to other building systems
According to expert	All types of experimental data	Material Product Building	Quick scan	Subjective results
Failure modes and effects analysis	All types of experimental data and developed data (obtained using other approaches)	Product Building	Easy to apply (apparently)	Subjective results

The analysis of this set of methods reveals the existence of sophisticated and interesting possibilities for assessing lifespans of materials, products and buildings. But how would these computer models have apprehended 16th or 17th century buildings or the large lifespan of the 18th century windows sometimes still in use? The behaviour of decision makers, who are influential in a consumer society and attached to the fast renewal of goods, could be a factor that would modify lifespans. In addition, the need to find more global solutions and take various, e.g. socio-economic, criteria into account has been pointed out (Baroghel-Bouny 2009).

1.2.2 Functional Lifespans of Products and Buildings

Sarja (2009) analyzed the studies and the tools dealing with the obsolescence and lifespan of buildings using technical principles. He stresses the urgent need to develop tools providing information on the lifespan and the limiting states of obsolescence, while noting that the issue is beyond the scope of the study of materials.

An analysis of the functional lifespan of a building focuses more on its social role and thus on how it responds to the perceived needs. The lifespan of a building can be affected by functional modifications, i.e. important changes in the needs that it can fulfil. In order to fully clarify the risks to lifespan, the approach proposed here is functional and systemic (Bertalanffy 1968). Using a technique based on the FIT® method, the various functions of the building are analyzed and evaluated. The building, fulfilling all the functions of use, is considered in its environment. After having listed the various elements of the system and named the associated functions, the approach consists of analyzing and evaluating the impact of the risks of functional degradations over the lifespan of the building.

For each set of features, it is necessary to ask, "What can make the need disappear? What can be changed? What is the extent of the risk?".

The functions of a building and the associated risks are presented in Table 1.3.

The reasons for modifying the lifespan are thus numerous.

Functions including a risk that can lead to demolition are:

- FI1—The building no longer to protects its occupants from the external environment because of the failure of non-exchangeable components, e.g. risk of collapse related to structural disorders or unstable ground. Nearby equipment with large nuisance effect. The risk of such an occurrence is relatively low.
- FI2—The building no longer allows social or professional activities to be carried out, e.g. disappearance of all professional, social or tourist activities, or does not promote links with the environment in which it is located, e.g. large developments that include no, or insufficient, integrated functions of a city.
- FA1—Occurrence of some events, such as an earthquake, can generate the end of the lifetime of the building.
- FA3—The absence of maintenance generates large disorders in the long term and thus the associated risk of demolition.

Table 1.3 Functional table—risk of obsolescence

References	Element of the system concerned by the function	Functions	Risks	Impact on building life
FI1	Occupants/Outdoor environment	Protect occupants from the external environment	Failure of exchangeable components	
			Failure of non-exchangeable components	X
FI2	Occupants/ Activities and environment	Allow occupants to perform their family, social and professional activities	Failure to adjust to new needs, change or absence of social and trade services	X
FA1	Climate	Resist external elements: wind, rain, snow, UV radiation, heat, natural disaster …	Failure of exchangeable components	
			Failure of non-exchangeable components	X
FA2	External environment	Reduce the environmental impact throughout the life cycle	Significant nuisance e.g. high energy consumption	
FA3	Maintainability/ Safety	Ease of maintenance	Lack of maintenance in the long term	X
FA4	Passers-by	Fit into and enrich the landscape, and flatter the eye of passers-by	Landscape changes and changing perceptions	
FA 5	Self-image	Match the image that the occupant has of himself and wishes others to have of him	Failure to correspond to a self-image	X
FA 6	Heritage/Real estate	Form part of heritage/real estate	Have no patrimonial value	X
FA 7	Culture	Form part of cultural heritage	Have no cultural value	

- FA5—A building that no longer corresponds to any "self-image" that an occupant may have, is doomed to demolition, especially in periods of low demographic pressure.
- FA6—The building no longer has any heritable or heritage value. While failure to fulfil a function of cultural heritage does not expose a building to a high risk of demolition, acquiring an element of cultural heritage status would protect it from such a risk.

Table 1.4 External risks that can lead to demolition

	Component	Functions	Risks	Impact on life building
C1	Rules/Policy	Not to be the subject of a demolition order	Being in opposition to the public interest	X
C2	Market	Resist the land pressure	Being in an area of high demand for land	X
C3	Climate	Resist violent climate phenomena	Damaged structure	X
C4	Xylophagous	Resist insect attacks	Being attacked by termites	
C5	Lignivorous	Resist fungal attack	Being attacked by fungi	
C6	Pollution	Resist pollution attacks	Degradation of wall cladding	
C7	Explosions	Resist external explosions	Damaged structure	X
C8	Earthquake	Resist earthquakes	Damaged structure	X
C9	Fire	Resist fire of external origin	Damaged structure	X

Regarding the principal function of FI1, protection, or FA1, interaction with the climate, the degradation of exchangeable components such as the outside wood finish, coatings, or cladding, do not have a direct impact on the lifespan of the building. Although the probability of the obsolescence or failure is high, these components can be exchanged or repaired. Exogenic constraints are likely to involve demolition. These constraints are presented in Table 1.4.

- C1: The building is the subject of demolition order. For example, the building occupies a space intended for a project of public interest, or is situated in a zone considered to present a risk for the occupants, or else does not perform the functions expected from such an element of the city any more;
- C2: The building is located in an area of significant pressure on land due to a large demand for housing;
- C3: The building is affected by violent weather or a climatic phenomenon involving its structure and/or putting its occupants in danger;
- C6, C7, C8 and C9: external phenomena such as an explosion, an earthquake or a fire can lead to demolition.

In summary, several causes associated with social functions can lead to demolition. It may be a question of, for example, a precautionary principle with respect to natural disasters, the disappearance of all local activities, the disappearance of any patrimonial value. The cause may also be the rejection by any occupant of the image reflected by the building, because of architectural mediocrity or the absence of maintenance. It may be that a building disrupts the coherence and the operation of a small island of the city. The end of lifetime can also result from an extraordinary accidental internal or external phenomenon. It should be noted that no thorough study appears to have been published on, and no tool exists for, the evaluation of these mechanisms in the building sector.

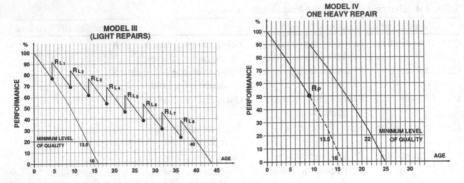

Fig. 1.1 Degradation model for a mineral plaster facade

1.3 Factors Influencing Lifespan

1.3.1 Impact of Maintenance on Lifespan

Many articles and works point out the importance of real estate management in terms of maintenance (Bonetto and Sauce 2006). Monitoring, maintenance, and repairs can, under certain conditions, extend the potential lifespan, as illustrated on the graphs below for the coating of a frontage (Flores-Colen and de Brito 2010). The authors of the article propose an optimization of the frequency of maintenance based on the systematic method (Fig. 1.1).

The issue is to extend the lifespan using preventive maintenance and to optimize the economics of lifespan related maintenance.

1.3.2 Maintenance and Maintenance Rules

Standards or regulations advising or requiring maintenance are rare, as is regular maintenance of building elements. Guides such as the FD P 05-102: 2003 guide contribute to the development of instructions for monitoring and maintenance of a house or similar construction. It enumerates the various works which could lead to such monitoring and maintenance by recommending a periodicity of the operations to be carried out. This document concerns all those involved in the construction of houses and in particular the building owners.

In addition to the specific thermal equipment, the elevators, the conduits of chimney and the VMC, maintenance or maintenance constraints are rare. Only the facades and the works containing wood are the subject of possible constraints or recommendations.

Sometimes, as in France, the facades are the subject of a specific regulation which can impose maintenance. This refers to articles 132-1 and following of the code of construction and the housing (dwelling) of the 31/12/2006 which is imposed upon Paris and municipalities which are included on a list established by decision of the administrative authority, on proposal or after assent of the Municipal councils. The article states: The facades of the buildings must constantly be held in good condition of cleanliness. Necessary work must be carried out at least once every ten years, on the injunction made to the owner by the municipal authority.

1.3.3 Adaptability

The concept of adaptability emerges at the intersection of the technical capacity of a building and its response to functional requirements. It is the ability of a building to adapt to changing needs in accordance with its lifespan. One of the factors in the durability of buildings is their operational flexibility or their adaptation to evolving functions. Examples: the constructions along the banks of the Garonne in Toulouse, France, originally built as storage and office spaces in the 18th century, were later turned into housing in the 19th and 20th centuries, or even offices, shops and housing in the 20th century. We could also consider this functional flexibility with respect to sustainable development, which would amount to asking the following questions: What were the economic, environmental and social costs of ensuring their standard of comfort and safety throughout the centuries? Langston et al. (2008) evaluated the gains, in terms of sustainable development, of reusing buildings. They showed that the potential gains of reusing a building are far from negligible. They cover various aspects:

- economical, since in general the cost of rehabilitation is significantly lower than that of demolition and reconstruction;
- temporal, considering the fact that renovation times are smaller than demolition/reconstruction times;
- environmental, on account of the efficient use of natural resources and the reduction of impacts;
- social, by the increased heritage value, as a result of their artisanal status.

In a stance promoting sustainable development, Kohler said that developed countries should focus on improving existing parks rather than build new buildings, as generally they merely replace the old (Kohler 1999).

1.3.4 Product Reuse and Recycling

Analysis with respect to life cycle has induced that the end of life be taken into account. This section presents the issues of waste and recycling in figures and details their impact on the ratings.

1.3.4.1 Statistical Measures

There seem to lack statistical data concerning waste generated at the level of the construction industry. Besides the difficulty of collecting information, the heterogeneity of specific technical solutions in each country probably explains this gap. For example, in France, the building sector accounts for about 15 % of the production of construction waste, i.e. 50 million tonnes per year (in comparison to about 30 million tonnes of household waste produced per year). 65 % of this waste comes from demolition, 28 % from rehabilitation and 7 % from new constructions. The construction sector is one of the main sectors affected by these nuisances (Source MEEDDM).[1]

Although recycling activity is not recent, "matter" (recycling) value of construction waste is estimated at less than 50 % (Source MEEDDM). Most of this is reused in the development of embankments.

Man has long been making recycling efforts and many ancient examples have been analyzed (Bernard and Dillmann 2008).

As regards recycling practices in France, according to Laurence Tubiana, Director of the Institute for Sustainable Development (Iddri),[2] "the construction and housing industry are lagging behind in terms of sustainable development." In the building sector, this is even worse: Peuportier evaluates it to 60 % for copper, 50 % for steel and 30–50 % for aluminum, lead and zinc (Peuportier 2001). It can be observed on the basis of our knowledge of current practices that most employed technical solutions are ill-disposed to recycling and at most allow for reuse in road construction.

1.3.4.2 Recycling Benefits

Through valuation models presented by Hendriks (Hendriks and Janssen 2003), the importance of taking recycling into account for assessing performance in terms of sustainable development is clearly established.

In 1995, only 50 % of waste was recycled in Japan. However, the energy saved by recycling aluminum may be 80 %, 40 % for steel and an average of 22 % for timber. Only the reuse of concrete results in an increase of about 5 %. By comparison of three examples, the traditional solution used for hardwood allowed for 67 % reuse of materials by weight (Emmanuel 2004; Arslan and Cosgun 2008). However, wooden frame is difficult to reuse because the connectors are scarcely removed without damage (Gao 2001).

In keeping with the terms of the life cycle analysis, the share of reused materials may be deducted either in the balance sheet of the initial construction or in that in which the materials are reused. However, the impact of the elements necessary

---

[1] MEEDDM—Ministère de l'écologie, du développement durable des transports et du lo.

[2] Institut du développement durable et des relations internationales.

for the removal operation must be added. Recycling could impose consequences on the life of recycled products as compared to the life of a new building. The phenomenon of decrease in lifetime must be considered. It is noted that the reuse of products does not seem to dramatically reduce the life of the new building. For example, the foundations of the lock house of Brienne in Toulouse does not appear to have reduced the life of this 230 year old building and is not slated for demolition as yet.

A brief look at the technical solutions used in France, however, shows that the recycling process is difficult to implement.

1.3.4.3 The Conditions for Recovery

Da Rocha and Sattler (2009) have studied the benefits of behavior change in the construction industry. The objective is to show the possibility of creating a more sustainable model output. The study focuses on the social, economic and legal factors that may act as obstacles or opportunities for reuse of building components. The results of the case study suggest that the reuse of building components is mainly due to the economic and social aspects such as the cost of deconstruction and demand for recycled products. Although legal mechanisms may play an important role in promoting the reuse of building components, factors influencing recovery are:

- Waiting time allowed for removal;
- The cost of deconstruction in relation to the application of the recovered products according to the purchasing power of the applicant;
- Monitoring and sanctions for non-compliance;
- The quality of the recovered products;
- Recovered material often associated with poverty in the minds of the population;
- Finally, the organization of the supply chain recovery.

Moreover, a very interesting study on the reuse of temporary housing has called for the development of a multi-criteria evaluation and has revealed the terms of the extension of use (Arslan and Cosgun 2008). It is therefore necessary to anticipate disasters and the organization of recycling items.

1.3.4.4 A Legal Framework, Objectives and Tools

France, in the European context and according to the Framework Directive of 19th November 2008 seeks to achieve its objectives by 2020. She wants to reach a minimum of 70 % reuse, recycling and material recovery of construction waste and demolition weight. The measures in the Grenelle Environment must achieve the target set by the Directive.

Recycling performance can be considered in the context of a comparison of different technical solutions. A study by Emmanuel (2004) in Sri Lanka establishes the result of a composite index based on material energy, overall cost and potential reuse. Several technical solutions wall are evaluated and compared. The study assesses the proportion of recoverable items for each of the technical solutions. The number of possible reuses ranges from 29 to 100 % even though they are not always performed for the same function (Emmanuel 2004). In this chapter, an index of ecological quality is developed based on three parameters: embodied energy, cost in terms of life cycles, and reuse. The study shows that material is ecologically the most efficient material while the least efficient is cement. However, the article concludes on the difficulty of using conflicting results. As a matter of fact, mud is not reusable. The article therefore concludes on the relativity of performance in terms of sustainable development and the inability to provide an overall performance score.

In summary, recycling is a direct factor influencing the performance of sustainable development. However, based on observations and considering actual technical difficulties, the construction industry still seems little invested in this practice, at least in some countries.

1.4   Tools for Assessing Performance

This section analyzes the consideration of lifespan in the different tools available.

1.4.1   Taking Account of Lifespan Using Standards

1.4.1.1   ISO 15686

Under ISO 15686—Buildings and constructed assets—Predicting lifespan, that of a construction product, is evaluated based on all available information, by multiplying the "standard lifespan" of the product by different correction factors. The objective is to take into account effects of the quality of the construction process from design, manufacturing and maintenance as well as internal and external stresses.

1.4.1.2   The French Standard NF P01-010

Intending to facilitate the designers' choice in environmental and health terms, the standard establishes a common basis for the issue of objective, qualitative and quantitative information. To facilitate comparisons of products, standard NF P01-010 introduces the concept of Functional Unit (FU). The functional unit is the function of a unit of product during one year. To assess the UF, the emission value recorded on the full life cycle is then divided by the "typical lifespan". The standard specifies that the typical lifespan should not be confused with the actual

Table 1.5 Typical lifespans mentioned in EPD

Product	Design office/ Manufacturer	Date	Typical lifespan (years)
Hollow concrete block	CERIB/Fédération de l'Industrie du Béton	September 2006	100
Monomur Terracotta 30 ground	Briques de France	August 2006	150[a]
Aerated concrete 25 cm	BIO IS/Syndicat nationale des fabricants de béton cellulaire	November 2007	100
Shuttered C25/30 CEM II with complex Ultra ThA thermoacoustic dubbing	ECOBILAN/Syndicat National des Bétons prêt à l'emploi	September 2007	100
Massive Noyant stone wall	CTMNC/LERM	July 2010	200
Glass wool	ECOBILAN/ISOVER	February 2007	50
Mineral plaster	ECOBILAN/SNMI	January 2007	50
Traditional wooden frame (05-027: 2009)		June 2009	100
OSB (Oriented Strand Board) N° 05-020: 2009		March 2009	100

[a]Between the beginning of our work during 2008 and February 2010, DDVT from 100 to 150 years after long negotiations

or theoretical length of the product life. It is representative of the product lifespan in the work studied for normal use and maintenance.

This estimate is announced and justified by the manufacturer on the basis of product usage guides. This data cannot guarantee the lifetime of the product, once implemented, to the extent that the manufacturer can not control the execution of the work in question. "Typical lifespans" identified in ESFD vary from 50 to 200 years. These times are shown in Table 1.5.

1.4.1.3 EUROCODE 0

Although its purpose is not to establish the lifespan, EUROCODE 0—The basis of structural design—implemented on 1 March 2010, defines the lifespans to take into account for structural dimensioning, depending on the nature of the building. These lifespan considerations are summarized in Table 1.6.

The «normal» lifespan in statistical terms for the structural dimensioning of a building is therefore 50 years.

1.4.2 Lifespan Consideration in Tools of Impact Assessment

Assessment tools require data. To obtain results with an appropriate level of reliability, it is important to assess data quality. With the intent of capitalizing on

Table 1.6 Lifespans for structural dimensioning of projects

	Approximate duration of use of project (years)	Examples
1	10	Temporary structures
2	10–25	Replaceable structural elements
3	15–30	Agricultural structures
4	50	Building structures and other common structures
5	100	Monumental structures and civil engineering projects

lifespan data, the Scientific and Technical Centre for Building (CSTB) is trying to develop a database named "EVAPerf" in the ISO 15686 standard format. Its purpose is to develop a collaborative platform providing information on the pathologies and lifespan of exploited products. The establishment of this database, very limited to date, is likely to take some time.

A study of various standards (Appendix 1) and tools for impact assessment (Appendix 2) in terms of sustainable development, has been developed. The method was initially to select repositories and tools that seemed most frequently cited in the articles and the most used in French research laboratories. The study, presented in tabular form, includes:

- A brief description to better understand the content and objectives;
- The strengths and weaknesses of each tool;
- Observations concerning lifespan consideration in the evaluation.

The study shows that most of the tools omit the consideration of lifespan in projects or existing buildings. The few to evoke this parameter, generally limit it to a few decades. When it is indeed taken into account, the building lifespans are fixed. Regarding the length taken into account, CASBEE and EQUER are the most ambitious tools with a length of 90 years. Within this maximum length, EQUER allows varying the lifespans of building components. By default, they are determined by the software on the basis of being defined "by an expert."

In summary, lifespans mentioned in the standards remain a matter of debate, considering the reasons for the retained values. These are even more important than presented as normalized values periods, and they could quickly become targets for professionals. Assessment tools currently do not allow for a serious consideration of the "lifespan" parameter.

1.5 General Analysis

1.5.1 The Window Example

Building lifespan is obviously related to the quality of the products used, the materials constituting these products, their implementation and maintenance. Taking up the window example, the tools developed seem scientifically indisputable. The calculated value is a range of 23–27 years. Yet aiming for a single main function, that of protecting the occupants from climatic stress and bring light in, the windows of old buildings in city centers may sometimes be more than two centuries old. Researchers and industrialists retort that the newer product is far more sophisticated because it performs more functions. It may also be cheaper and many windows of lesser quality have had a shorter lifespan. We have no statistical information about this. In any case, a preliminary observation shows that this is not necessarily the technical lifetime that corresponds to lifespan, but other refers to other criteria of a functional nature. A large number of windows have been scrapped due to lack of response to new features such as thermal or acoustic insulation. These are the causes of obsolescence. It is also possible that in some cases the expiration of a product is purely psychological, pertaining to its "associated" product value. Dupuy and Gerin (1975) describe this phenomenon in their book "De la société du toujours plus à l'obsolescence psychologique des biens". The ambitious program EVAPerf will face great difficulties in incorporating these factors to measure the lifetime of products. Another issue could be the consumer's interest and willingness to remedy the deficiencies that led to disposal. If such a "repairability" is envisaged during the design process, would be its production cost? Its environmental and social costs? What would be its impact on the industrial and commercial sectors?

1.5.2 Recycling and Recovery

Recycling or recovery of products constituting the building are factors that favor lifetime extension in a certain way, even if their function is changed.

1.5.3 EUROCODE 0

It is observed that the technical lifetime of a building is related to the lifespan of its foundations and load-bearing walls. What results would the assessment tools presented here yield, when applied to these structural elements? There are numerous old buildings on which they could be applied. In 2006, in France, the INSEE[3] accounted for 12 million housings over 50 years old and 5.33 million over a 100. These large numbers cannot merely be the result of chance. What lifetime would these new assessment tools predict for these structures? How did EUROCODE 0 determine a value of 50 years for the assessment of the newer structures? Does this value reflect the accepted lifespan used in other countries? Or is it the result of a collective will of the society? Does it result from optimized economic, environmental

---

[3] INSEE: 'Institut national de la statistique et des études économiques' or National Institute of Statistical and Economical Studies collects, produces, analyses and distributes information about the French economy.

and social costs, or is it an approach towards sustainable development? Wouldn't this standard lead our constructors and product manufacturers to adopt a similar lifespan for their products? In the latter case, since this would amount to the lifespan of a loan, would it be satisfactory to consumers? It is quite probably not based on an analysis of the needs of the client or the user. Even so, it is likely that this term will became a standard for professionals.

This contradiction between the probable requirements of users in need of housing, undoubtedly on a long-term basis, and the established standard which sets a short lifespan for buildings, is remarkable. It is especially so when compared to the problems of life in the realm of living things. Zaoui wrote "Yet, masters of development and design, free of any additional bioethical concerns or suspicion of eugenics, the researcher and material engineer, in contrast to the biologist or the doctor, have the flexibility to change, at will, the conditions and rates of aging, and use mechanisms to increase their 'wealth'" (Zaoui 2002). We should also be able to take action on the lifetime of buildings.

1.5.4 Concerning Maintenance

Maintenance is a decisive factor since it is responsible for the conservation of the functional characteristics of products as well as the building itself. Its impact has not been studied in depth since there are too many products, and too many factors of influence. In fact, maintenance recommendations are related to the original qualities and usage of products. They are also related to environmental and building usage conditions. Determining maintenance optimization factors for lifetime extension is therefore quite problematic. Furthermore, the lifetime of a building is not equal to the aggregate of the lifetimes of the products it consists of. On the same principle as in the window example, building lifespan analysis shows that impact factors can be associated with commonly used functions.

1.5.5 Adaptability

In addition, the more a building is flexible in its operation, more it can adapt to changing needs. This refers directly to the quality of the designer's work and reduces the risk of demolition. It is called a functional "recycling" of the building. However it seems that few methodological design or assessment tools have been developed for this criterion.

1.5.6 Optimization

Whether term considered is technical lifespan or functional lifespan, the design decisions concerning optimization should make use of tools such as value analysis. What technical options will lead to the anticipated results for the desired length of time at the lowest overall cost or even at the lowest extended overall cost? This implies having the necessary knowledge of the expected needs and a forecast of their development, or the management of random phenomena. When building products have a lifetime greater than the function they are supposed to fulfill, value analysis demonstrates the default optimization. Indeed, consumed material, human and economic resources are higher than they have been. In contrast, insufficient means do not imply optimization. In fact, the reconstruction necessary to meet the public use on the rest of the "time remaining" may fall within the scope of optimization.

1.5.7 Need

In addition to these aspects of the analysis, the issue of needs should be central to the problem of lifespan. The aim of the "builders" of the past was patrimonial transmission to subsequent generations. The idea was to make life easier for future generations, allowing them to devote their resources to other needs. The aim of owners and managers of goods or service production companies are to match usage lifetime with that of tax depreciation or at least that of the gross operating profit. Finally, the aim might be that of optimizing the lifetime for sustainable development. This objective takes into account social, environmental and economic cross factors. For example, it is shown that the demolition of a building presents a social coup for occupants, who now have lost all bearings to their past life (Rojas Arias 2007). This cost, assuredly difficult to assess, is not included in current studies.

1.6 Conclusion

The technical lifetime of products and buildings is the object of studies for developing assessment tools. The question of the functional life of the buildings is treated less extensively.

About the "social" causes of end of life, a part of the responsibility lies with the designers. In fact, they are responsible for a building's adaptability to changing needs as well as its "maintainability". A thorough analysis of the characteristics of adaptability to changing needs seems necessary for durability, and deeper analysis of the causes of the functional aging associated with human behavior would probably shed further light on the subject. Moreover, designers are also responsible for cost optimization in the broad sense when making technical choices. However, they lack the tools to take decisions in terms of this optimization.

Owners and users also have their share of responsibility. Building usage and maintenance along with their personal values and aesthetic sensibility are all factors that contribute to its lifespan.

The existence of numerous examples of centuries-old buildings suggests that they cannot be the result of mere chance. It remains however to be considered whether they have been partially or completely rebuilt or not. The archives of the Waterways of France (VNF) buildings attest to the almost complete conservation of an important 18th century property simply on the basis of careful supervision and minimal maintenance. Thanks to these records it is now possible to establish a maintenance booklet. Without knowing the end of a building's lifetime, we could evaluate its performance in terms of sustainable development.

In summary, the actual life of a building is generally the shorter amongst the technical or functional life of the building. The technical life is linked to product quality and the functional life depends on owner and user care and practices. Maintenance is a common factor which may alter both lifespans. Finally, other exogenous factors could be the cause of demolition.

Lifespan is a variable upon which we can act!

References

Arslan H, Cosgun N (2008) Reuse and recycle potentials of the temporary houses after occupancy: example of Duzce, Turkey. Build Environ 43(5):702–709

Baroghel-Bouny V (2009) PLATEFORME OUVRAGES D'ART, Développement d'une approche globale, performantielle et prédictive de la durabilité des structures en béton (armé) sur la base d'indicateurs de durabilité. Publication LCPC http://www. piles.setra.equipement.gouv.fr/article.php3?id_article=564

Bernard JF, Dillmann P (2008) Il reimpiego in architettura: recupero, trasformazione, uso. Collection de l'École française de Rome, No 418

von Bertalanffy L (1968) Théorie générale des systèmes. 2e éd. Dunod, 25 septembre 2002

Bonetto R, Sauce G (2006) Gestion de patrimoine immobilier—Les activités de références CSTB. Département Technologies de l'Information et Diffusion du Savoir, Université de Savoie, Polytech'Savoie—LOCIE

Choay F (sous la direction), Merlin P (2010) Dictionnaire de L'urbanisme et de L'aménagement, 3rd ed. Presses Universitaires de France—PUF

Cooper T (2004) Inadequate life? Evidence of consumer attitudes to product obsolescence. J Consum Policy 27: 421–449 (The Netherlands)

Da Rocha CG, Sattler MA (2009) A discussion on the reuse of building components in Brazil: an analysis of major social, economical and legal factors. Resour Conserv Recycl 54(2):104–112

Dupuy JP, Gerin F (1975) Societe Industrielle et Durabilite Des Biens de Consommation. Revue Economique 26(3):410–446

Emmanuel R (2004) (Vogtländer JG, Hendriks CF, Brezet HC 2001) The EVR model for sustainability—a tool to optimise product design and resolve strategic dilemmas. J Sustainable Prod Des 1(2): 103–116

Flores-Colen I, de Brito J (2010) A systematic approach for maintenance budgeting of buildings façades based on predictive and preventive strategies. Constr Build Mater 24(9):1718–1729. doi:10.1016/j.conbuildmat.2010.02.017

Gao W, Ariyama T, Ojima T, Meier A (2001) « Energy impacts of recycling disassembly material in residential buildings. » Energy Build 33(6):553–562

Hendriks CF, Janssen GMT (2003) Use of recycled materials in constructions. Mater Struct 36(9):604–608

Kohler N (1999) The relevance of the green building challenge: an observer's perspective. Build Res Inf 27(4/5):309–320

Lair J et al (2003) Ingénierie du développement durable: Vers la formalisation d'une doctrine française. CSTB, Département XXIEMES RENCONTRES UNIVERSITAIRES DE GENIE CIVIL 2003

Langston C, Wong FKW, Hui E, Shen LY (2008) Strategic assessment of building adaptive reuse opportunities in Hong Kong. Build Environ 43(10):1709–1718

Lemer AC (1996) Infrastructure obsolescence and design service life. J Infrastruct Syst 2(4):153–161

Levy J-P, Saint Raymond O (1992) Profession propriétaire. Logiques patrimoniales et logement locatif en France, Toulouse, PUM, collection "État des lieux", 1992

Miller R (2002) Biologie du vieillissement lettre no 3/printemps2002 de l'Académie des sciences

Peuportier B (2001) Training for renovated energy efficient social housing—section 2 tools. In: Intelligent energy European program, no EIE/05/110/SI2.420021

Rojas Arias JC (2007) Thèse La politique de la démolition: Rénovation urbaine et habitat social en France et en Colombie

Sarja A (2009) Reliability principles, methodology and methods for lifetime design. Mater Struct 43(1-2):261–271. doi:10.1617/s11527-009-9486-y

Talon A (2006a) Evaluation des scénarii de dégradation des produits de construction. Thèse Génie Civil. Clermont-Ferrand: Centre Scientifique et Technique du Bâtiment—service Matériaux et Laboratoire d'Etudes et de Recherches en MEcaniques des Structures, 2006, 240 p

Talon A (2006b) Evaluation des scénarii de dégradation des produits de construction. PhD dissertation, Université Blaise Pascal Clermont II, France

Wang KM, Lorente S, Bejan A (2006) Vascularized networks with two optimized channel sizes. J Phys D 39:3086–3096

Zaoui A (2002) Les matériaux vieillissent aussi. lettre no 3/printemps2002 de l'Académie des sciences

Chapter 2
Building and Sustainable Development

Abstract This section explains the concept of sustainable development in terms of a building. Multiple sources for environmental indices imply the need for a thorough qualitative analysis on the basis of established criteria. The first part lays the background and the second specifies the indicators and associated data to measure performance. The next part presents different assessment tools or guides that improve performance. Finally, the fourth and last part presents and analyses the scientific literature.

Keywords Building • Tools • Sustainable development • Assessment

This section explains the concept of sustainable development in terms of a building. The first part lays the background and the second specifies the indicators and associated data to measure performance. The next part presents different assessment tools or guides that improve performance. And finally, the fourth and last part presents and analyses the scientific literature.

2.1 Background

The importance and impact of the negative effects of the growth of industrial activities at a global level have led to the development of the concept of "sustainable development". It has rapidly become a long-term concern for buildings. There exist several definitions of sustainable development. While not opposing and often complementary, these definitions are vague in nature, especially when the analysis focuses on the consequences they entail, both at individual and collective levels.

M. Méquignon and H. Ait Haddou, *Lifetime Environmental Impact of Buildings*,
SpringerBriefs in Applied Sciences and Technology, DOI: 10.1007/978-3-319-06641-7_2,
© The Author(s) 2014

Fig. 2.1 Model Jacobs and
Sadler (Boothroyd 1990)

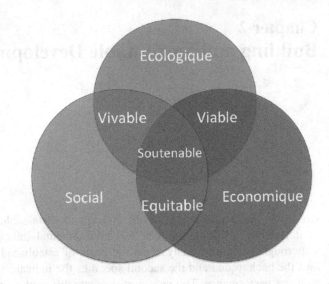

2.1.1 Definitions

The following relatively accurate definition allows us to address the essential
issue of sustainable development. It was put forth by the World Commission on
Environment and Development:

"Sustainable development seeks to meet the development needs of present generations without compromising the ability of future generations to meet their own
need" (WCED 1987). This definition seems to have a relative consensus among
most of the people interested in the subject (Fig. 2.1).

At the Copenhagen summit in 1995, it was recognized that sustainable development should be characterized by three dimensions that would give it a fuller and more
accurate content. These are social, economic and ecological dimensions. This charac-
terization is illustrated by the above model of Jacobs and Sadler (Boothroyd 1990).

The originality of sustainable development lies in the systemic or structural
connection between economy, the environment and society. It is understood as a
development that meets the basic needs of one generation without compromising
the needs of other generations.

The economic aspect reflects the search for sustainable development focus-
sing on growth and economic efficiency. This approach must address the need to
develop the economy and the society, particularly in the case of developing countries seeking an adequate standard of living.

The social approach expresses the fact that sustainable development must be
based on human needs and therefore aim for social equity. The individual positioned
at the centre of the action can meet this necessity. Recalling the intra-generational
and intergenerational links, the Brundtland Report has positioned man at the centre
of the objective; this approach involves health, hygiene and cultural aspects. In terms
of the intergenerational aspect, the report also set the goal in relation to time.

The ecological approach stresses the fact that action has to preserve, improve and enhance the environment and make it fit for life. Action must conserve resources on the long term and favour regeneration rather than exhaustion. The approach also includes the reduction of climate impacts caused by human actions.

The different features of the concept are:

- A relationship between the environmental, social and economic aspects;
- A transverse and systemic approach;
- Harmonization between short term and long term, based on the precautionary principle;
- A motto of "think global, act local";
- Solidarity between rich and poor countries, with an inter-generational component;
- A new form of governance for strengthening democracy.

The question of the role of culture arises through the definition. Can we ignore the approach to issues of the built environment? In its Opinion No. 2002–2007 of April 2002, the French CSD questioned the lack of reference to culture in the work on sustainable development. In this text, the Committee puts man at the centre of the device and reminds the specificity of the species in its relation to culture. She emphasizes "the need to complete the approach to sustainable development by integrating the cultural dimension as well as economic, social and ecological dimensions." The actions undertaken in the framework of a sustainable development approach must necessarily integrate the cultural specificities of each human group. Cultural diversity, as well as natural heritage must be protected and enhanced in order to be transmitted to future generations.

2.1.2 Principle: Analysis of Life Cycle

According to ISO 14040:2006 standards Series: Environmental management—Life cycle assessment—Principles and framework, the analysis lifecycle or "LCA" assesses the environmental impact of a product or system considering all stages of its life cycle.

The objective of this principle allows, firstly, to identify the points on which a system can be improved and secondly, to obtain a complete record of the essential part of a performance comparison of different solutions. According to the attached diagram extracted from the ISO 14040 standard, the method consists of four main phases (Fig. 2.2):

These are:

- Define the objectives and define the system;
- Create the emissions inventory. This is the quantitative description of the different flows that cross system boundaries. The inventory should consider recycled or reused elements and shall be deducted from the global value;

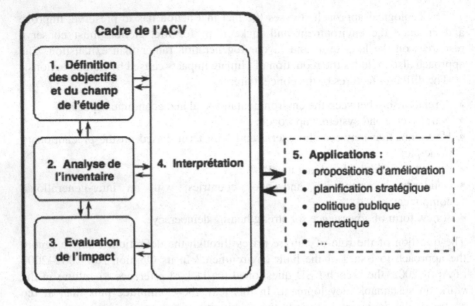

Fig. 2.2 ACV steps—ISO 14040-2006

- Make an interpretation;
- Analyse the impact: they may fall within human toxicity, noise, creating oxidant, the depletion of the ozone layer, global warming, acidification, eutrophication, eco-toxicity, land use and loss of habitat, dispersion of species and organisms, the usage of natural resources, soil erosion, salinization of soils, … (Curran 2006).

In brief, products or systems must be evaluated in their life cycle, which means taking into account the phases of design, production, use and demolition. Only the life cycle analysis of products used to compare different solutions in the context of an assessment of environmental, economic and social impact.

2.1.3 Eco-Design and Eco-Building Materials

Architecture compatible with sustainable development therefore results from a subtle balance between the requirements of the society, economy and environment.

The building includes several functions to meet many needs. The functional response can be established through numerous solutions. These responses are established at a level of environmental, social and economic quality. The study of ecological quality is the most thorough approach. The environmental quality of buildings includes the quality of indoor environments and reducing impacts on the environment. Eco-design includes the choice of materials and the system consisting of assembly, as regards the impact they induce on the inside and outside environment. This analysis assesses the degree of performance of the choice of materials as eco-materials (Peuportier 2001a, b).

2.1.3.1 Definitions

Eco-materials are materials classified according to their environmental and health impact, according to their performance and comfort based on the lowest overall cost. Therefore, the materials are classed by their performance levels.

Eco-design of a building take into account the environmental impact and atmosphere through the morphology and organization of interior spaces for the lowest overall cost (Peuportier 2001a).

2.1.3.2 The Criteria for Eco-Materials

The criteria mentioned in the French standard NF P01-010 impacts the choice of eco-materials are introduced implies that performance in the following terms (Table 2.1):

Table 2.1 Eco-materials properties

Specifications	Units	Definition (Sacadura 1993)
Density	–	Ratio of the mass of material and the mass of same volume of water at the temperature of 3,98 °C.
Thermal Conductivity	W/m.K	Heat flow in watts running through thick 1 m materials on a surface of 1 m^2 with a temperature of 1 °C or K between the two sides
Thermal capacity	$Wh/m^{3*}K$	Ability of the material stored heat. It measures the amount of heat required to raise 1 °C, 1 m^3 of material.
Phase shift	h	Speed of the heat wave to pass through a material
Thermal effusivity	$J^*K.m^{-2}.s^{-1/2}$	Coefficient that characterizes the speed with which the temperature of a material is heated
Thermal diffusivity	m^2/s	Physical quantity which characterizes the ability of a material in the penetration and the alleviation of a thermal wave in a medium
Porosity	–	Ratio of the volumes of voids on the volume of the materials
Hygroscopic	%	Ability to hold water and interact with the environment
Sound reduction	dB	Ability to absorb sound waves

2.1.3.3 The Criteria for Eco-Design

The impact criteria mentioned in the previous paragraph, an eco-design will aggregate other analysis such points:

- Land use;
- Surrounding cover;
- The winds;
- Sunlight causing overheating in summer and glare;
- Ambient light that will change the lighting and energy, and liveability;

- The luminance of a light source is the ratio of the intensity of the source on a surface in a direction of the projected area of the source;
- Colors and surface conditions;
- Noise;
- Air, its renewal, its speed and relative humidity;
- The energy sensors;
- Leakage and radiation.

Bernstein et al. (2006), Liébard and De Herde (2006), Déoux and Déoux (2004).

The response to the program of the client, must be a consensus in the consideration of various constraints so that the impact on the environment is as low as possible and the indoor environment most habitable. This optimal response must result from an equilibrium having a minimum overall cost.

Finally, as recommended by the 2nd target of the 2008 version of the HQE®, Bruno Peuportier evokes the "sustainability" within the meaning of lifespan in his general analysis of buildings (Peuportier 2001b). It presents rehabilitation of old buildings obtaining a very satisfactory performance in terms of energy but does not evaluate the gains in other criterion of sustainable development nor those obtained by extending the lifespan. Lifespan of products, components built using these products and building itself has an impact on the environment. None of the other works that have been consulted involve the question of the lifespan of products or buildings and their influence in performance.

2.2 Indicators and Sustainable Development Data

2.2.1 Indicators and Indices

In this section, indicators and indices are defined in the sense of Boulanger (2004). Thus, an indicator is an observable variable used to reflect the status of a non-observable reality. For example, it may be the amount of greenhouse gas emitted by the manufacture of a product, the unit being kg of CO_2. An index refers to a synthetic indicator constructed by aggregating so-called basic indicators.

The process of construction of indicators is identified by Lazarsfeld in 1958 according to the following scheme:

Du concept aux indices

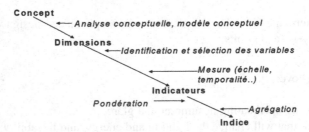

Table 2.2 List of polluting emissions

Consumption of non-energy natural resources

Antimony (Sb), Silver (Ag), Clay, Arsenic (As), Bauxite (Al_2O_3), Bentonite, Bismuth (Bi), boron
  (B), Cadmium (Cd), Limestone, Sodium Carbonate (Na_2CO_3), Chloride potasium (KCl),
  Sodium chloride (NaCl), Chromium (Cr), Cobalt (Co), Copper (Cu), Dolomite, Tin (Sn),
  Feldspar, Iron (Fe), Fluorite (CaF_2), Gravel, Lithium (Li), Kaolin (Al_2O_3, $2SiO_2$, $2H_2O$),
  Magnesium (Mg), Manganese (Mn), Mercury (Hg), Molybdenum (Mo), Nickel (Ni), Gold
  (Au), Palladium (Pd), Platinum (Pt), Lead (Pb), Rhodium (Rh), Rutile (TiO_2), Sand, silica
  (SiO_2), Sulfur (S), Barium sulfate ($BaSO_4$), Titanium (Ti), Tungsten (W), Vanadium (V) Zinc
  (Zn), Zirconium (Zr)

Emissions in air

GHG and acidification Nitrogen oxides

GHG: Carbon dioxide (CO_2), Methane (CH_4), Hydrofluorocarbons (HFCs), Perfluorocarbons
  (PFCs), Sulfur hexafluoride (SF_6)

Sulfur dioxide

Greenhouse eutrophication Ammonia (NH_3)

Hydrocarbons, polycyclic aromatic hydrocarbons (PAHs), volatile organic compounds (VOC),
  Carbon monoxide (CO), Nitrous oxide (N_2O), Sulfur oxides, Sulfur Hydrogen, Hydrogen
  cyanide acid, chlorinated organic compounds, inorganic chlorine compounds, chlorinated
  compounds not specified, fluorinated organic compounds, inorganic fluorine compounds,
  halogenated compounds, fluorine compounds unspecified Cadmium and its compounds,
  Chromium and its compounds, Cobalt and its compounds, Copper, Tin and its compounds,
  Manganese, Mercury, Nickel and its compounds, lead and its compounds, Selenium,
  Tellurium Zinc, Vanadium Silicon

Emissions in water

COD (Chemical Oxygen Demand), BOD5 (Biochemical Oxygen Demand in 5 days), Suspended
  Matter (SPM) AOX (adsorbable organic halogens compounds), hydrocarbons, phosphorus
  compounds

This aggregation of heterogeneous indicators requires a multi-criteria approach
which is made to establish conversions to homogeneous units on the basis of the
chosen relative importance.

The list of product emissions in the building sector is accurate cf. Table 2.2. It
allowed the establishment of estimating emissions of products based on the French
standard NF P01-010, itself supported on ISO 14040 series of analytical life cycle
(Life Cycle Impact Assessment). The objective is the evaluation of an entire life
cycle, from product emissions in the building sector.

In these programs, you must add the evaluation of water consumption and the
inventory of recycled materials.

On the depletion of natural resources, the index used is the Abiotic Depletion
Potential (ADP). This indicator takes into account the consumption of energy and
non-energy resources (except water) by weighting each resource by a coefficient
corresponding to a Rarity (antimony has a value of 1 by convention) (Clift 2004).
In ESFD base INIES, a value greater than 1 for UF of a product indicates that we
are consuming a resource that's rarer than antimony. The resources whose value
on the indicator is very low (less than 0.001) are considered non-exhaustible on a
human scale. This principle is stated in the standard NF P01-010.

Fig. 2.3 Diagram of
constituting an index
of environmental impact
(Oberg 2005)

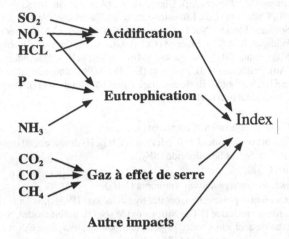

The formula for calculating the depletion of natural resources is as follows
(Habert et al. 2010a, b):

With DRi (kg/year) extraction rate of resource i

Drsb is the extraction rate of antimony which is equal to 6.06×10^7 kg/year
and Ri is the stock of resource i in kg

RSb equal to 4.63×1015 natural resources stock of antimony (Sb).

For example: The maximum stock of cement in France

$$R \text{ cement} = 2.49 \times 1012 \text{ kg}$$

$$DR \text{ cement} = 2.48 \times 1010 \text{ kg/year}$$

$$ADP \text{ cement} = 1.41 \times 10^9 \text{ kg eq Sb}$$

In conclusion of this example, the calculation performed for UF, is well below 1.
These results present the cement as weakly exhaustible. Nevertheless, discussions
and oppositions may appear on the available resources to the extent that the rate of
extraction and resources are located.

On the establishment of an index globalising environmental impacts, work has
been completed. The logic is shown schematically in the figure below provided by
Oberg (2005), Osso (1996) (Fig. 2.3).

Then, the difficulty lies in choosing the relative impacts. On the basis of the
previous diagram, proposed constitution of indices were performed as illustrated
in the Table 2.3.

One can sense the difficulty of consensus in the rapid analysis of proposed dis-
tributions of impacts. For example, contradictions are evident in the proportions
of nitrogen compounds and sulphur dioxide in SIKA and EPS systems. Also,
only a few impacts such air and energy used are taken into account. For example,
impacts such as water emissions or resource depletion are omitted. Integration of

Table 2.3 Proposed key distribution impacts (Oberg 2005)

Sytem	Units	Emission in air			Energy used	
		CO_2	NO_x	SO_2	Fossil	Electricity
SIKA[a]	Euro	0.088	6.8	2.3		
EPS[b]	ELU	0.108	2.1	3.3	0.0094	0.0009
ET[c]	–	36.5	3,970	3,770	2.94	2.78

[a]SIKA (2002)
[b]EPS 2000 (Stenn 1999)
[c]Baumann and Tillmann (2004)

all impacts, would require a consensus on these impact factors, their consequences and the importance of their relative impact.

In summary, on environmental indicators, there is no consensus on the creation of an index. The lack of consensus is not about impact indicators taken into account but the relativity of impacts between themselves. The aggregation of indicators was not successful and, in fact, the production of a consensus index is not performed.

2.2.2   Emissions from the Sector and Its Products

Ecological context includes many factors. The high profile highlighting the risks of climate change and depletion of natural resources, is not new. The risk of global warming associated with the greenhouse effect had been identified by Arrhénus in 1896 (Dufresne et al. 2006). The fear of depletion of coal resources was raised by Jevons in (1866). However, consumption and impacts resulting from rapid population growth, have imposed awareness. So at the Earth Summit in Rio de Janeiro, Brazil that the international community has truly become aware of the issue in terms of global warming, climate change, resource depletion, and destruction of flora and fauna.

Although vitiated by many uncertainties (Lorius 2003; Le Treut et al. 2008) and sometimes even disputes (Enghoff and Svensmark 2008), a number of climatic effects are accepted by the entire scientific community. In 1999, Michel Petit listed, changes in temperature, atmospheric carbon dioxide concentration, the influence of water vapour and rising sea level as certifiable phenomena.

At international level, the United Nations established in 1988, the Intergovernmental Panel on Climate Change (IPCC), whose mission is to study the issue of climate change. While in its first report in 1988, the IPCC still hesitated to hold man responsible for global warming, the fourth report in 2007 leaves no doubt on the issue.

For example, in France, the building sector is the largest consumer of energy among all economic sectors, with the equivalent of 65.35 million tonnes of oil in 2009, or 43.88 % of the total final energy (Source: energy Statistics France, March 2005—Ministry of Ecology, Sustainable Development, Transport and Housing—all metropolitan areas).

This energy consumed annually results in the emission of 120 million tonnes of CO_2 (Source: Inventory of air pollutant emissions in France, February 2010)—CITEPA (Inter-professional Technical Centre for the Study of Air Pollution), representing between 23 and 25 % of national emissions (Source ADEME: Agency for Environment and Energy Management). It is still responsible for 466 million tons of minerals extracted for construction which represents nearly 75 % of global consumption ("stats.environnement.developpement-outenable.gouv.fr"). Regarding waste, the construction sector accounted for 343 million tonnes in 2004 (Source ADEME—waste figures) which is the largest generator of waste, ahead of household garbage with 26 million tonnes.

The construction of buildings is responsible for several important impacts on the environment. Through the scientific literature, a consensus about the causes of environmental impacts has taken place. The main pillars of ecological field to be evaluated are set within ISO 14044: 2006 Environmental management—Life cycle assessment—Requirements and guidelines. Pollutants to assess relate to acidification, eutrophication, photochemical pollution, greenhouse gas emissions, contamination by heavy metals, contamination by persistent organic pollutants and suspended particles.

CITEPA (Inter-professional Technical Centre for the Study of atmospheric pollution) is responsible for the implementation of the emissions inventory in France.

It follows that the only residential and tertiary sectors in France are the source of air emissions in 2008 in the following proportions (Source: CITEPA/CORALIE/FORMAT SECTEN—Update April 2010) (Table 2.4).

Note that for many emissions, significant percentages often apply to quantities in sharp decline since the 1990s. For example, SO_2 emissions decreased by 74 %, 63 % PAHs or PCBs 28 % for all sectors. However, a large part of the emissions are generated by the residential sector. Improved technologies using more biomass explains the overall reduction in emissions from the sector. These reductions in emissions are important because they apply to an annual production increase of nearly 45 % of housing between 1990 and 2007. Housing units have a surface increase of nearly 6 % over the period (Source INSEE) These "structural" changes in the sector, combined with the increase of 7.3 % of the population, largely explain the 22 % increase in greenhouse gas emissions measured between 1990 and 2004.

At the scale of the building itself and on the impacts described above, the phases of the life cycle do not have the same magnitudes in the impact assessment. For example, the work of researchers such as Liu and Pulselli showed that at least in some areas of the building in terms of energy consumption and environmental impact, the usage phase of the building was much larger than construction, maintenance and demolition. However, large differences appear from one item to another. Mr. Liu shows that 70–80 % of the energy and environmental impacts are attached to the use of 50-year phase (Liu et al. 2010) while Pulselli estimated at 49 %, the manufacturing phase, 35 % of the maintenance and 15 % using (Pulselli et al. 2007). The proportions of impacts on which these statements are based probably see some of their differences because of the different places the studies have taken place, the city of Chongqing in China for the first, Italy for the second.

Table 2.4 Emissions from the residential and tertiary sectors in 2008

Substances causing acidification, eutrophication, photochemical pollution

9 % du SO_2	8 % du NO_x	31 % de COVNM	32 % du CO

Substances for increasing the greenhouse gas

23% of carbon dioxide (CO_2)	3 % de methane (CH_4)	2 % of nitrous oxide (N_2O)	1 % de sulfur hexafluoride (SF_6)
		52 % des hydrofluorocarbons (HFCs)	

Substances for contamination by heavy metals

15 % of arsenic (As)	6 % cadmium (Cd)	25 % of chromium (Cr)	3 % copper (Cu)
7 % of nickel (Ni)	15 % of lead (Pb)	9 % dcf selenium (Se)	25 % of zinc (Zn)
			5 % of mercury (Hg)

Substances related to contamination by persistent organic pollutants

17 % of dioxins and furans (PCDD-F)	68 % of polycyclic aromatic hydrocarbons (PAHs)	20 % of polychlorinated biphenyls (PCBs)
		20 % of polychlorinated biphenyls (PCBs)

Suspended particles

10 % des TSP	60 % de $PM_{1,0}$	34 % de $PM_{2,5}$	22 % de PM_{10}
	60 % de PM_{10}		

Table 2.5 List of groups of environmental impacts in the building

Total primary energy (MJ)	Renewable energy (MJ)	Non-renewal energy (MJ)
Primary energy process (MJ)	GHG (kg éq CO_2)	Reused waste (kg)
Hazardous waste (kg)	Non-hazardous waste (kg)	Inert waste (kg)
Radioactive waste (kg)	Air pollution ($m^{3)}$	Water pollution ($m^{3)}$
Acidification (kg éq SO_2)	Resource depletion (sb)	Water (L)
Photochemical ozone formation (kg ethylene eq)		
Destruction of the stratospheric ozone layer (kg eq)		

This observation implies that it is necessary to distinguish the impact of different phases, for the comparison of technical solutions which take into consideration the issue of lifespan.

On the scale of construction products, Standard NF P01-010 (NF P01-010-2004 Environmental quality of construction products—Environmental Statement and health of construction products), established from the ISO 14040 series of standards (ISO 14040—Environmental Management—life Cycle analysis) based on the principle of life-cycle analysis, and taking all of the inventory prepared by the ISO 14025 standard (ISO 14025—environmental labels and declarations—Type III environmental declarations) led to the establishment of reporting environmental and health records (EPD). This Standard applies to the presentation of different impacts of products used in the building. Different impacts are grouped under the headings of the following indicators whose definitions are provided in the Table 2.5.

Work on the environmental impact of human activities mainly address the issue of climate change. On the depletion of natural resources, we take the Life Cycle Impact Assessment Method (LCIA) develops an indicator for resource depletion (Habert et al. 2010a, b). For the measurement, the principle applied is that of the pressure on stocks of antimony. However, G. Habert highlights the importance of the local level to take into account the depletion of natural resources. Based on a study of the Paris region, it shows how the availability of resources can be calculated and attempts to measure the forecast error associated with economic or technologically complex social situations (Habert et al. 2010a, b).

In summary, the DU10 indicators seem to be a standard for the evaluation and measurement of the environmental impacts of an anthropogenic origin.

2.2.3 The Characteristics of the Data and Data Sources

Evaluate performance and simulate behaviours implies the use of data. On environmental indices of products used in the building, some organizations have undertaken the creation of databases. The most serious and known organisations publish the specific protocols for the preparation of the databases. These databases are listed and briefly analyzed in the Table 2.6.

Table 2.6 Database of sustainable development

	Sectors	Nature et specialities	Qualities	Défects
ECOINVENT Switzerland www.ecoinvent.org	All	Database co-managed by government agencies and professional / Environmental database of materials and systems for the building	Tool inventory comprehensive information lifecycle data global warming, acidification, primary energy, renewable, non-renewable, eutrophication / To facilitate environmental claims of products, life cycle analysis, the management lifecycle and design for environmental / Very often used in assessment tools	The high cost of license
KBOB Switzerland www.bbl.admin.ch	Building	Coordination conference construction services and building master public authority KBOB	Database which provides for the elements constituting the primary energy consumption and GHG emissions of the solutions building / Evaluation based on data ECOINVENT / Free	Do not provide information on primary energy and GHG
«Baubook» Austria www.baubook.at/zentrale	Materials and building system	Energy		Source of data and protocol difficult to obtain
«ELCD Database» EU http://lct.jrc.ec.europa.eu	All	Database co-managed by government agencies and professional / Environmental impacts	Willingness to harmonize the index values of environmental impacts of products across Europe / Free database	Base in preparation poorly documented / Data from professional associations whose control is difficult to implement / Go to face the opposition of approach to wood-based materials

(continued)

Table 2.6 (continued)

	Sectors	Nature et specialities	Qualities	Défects
INIES France www.inies.fr	Building	Database co-managed by government agencies and professional Environmental and health impacts	Based on the NF P01-010 standards established itself on the ISO 14040 series for inventory evaluation in life cycle and ISO 14020 for environmental labelling Provides data for global warming, acidification, primary energy, renewable, non-renewable, eutrophication Operation based on a specific protocol between government agencies and professionals under the control of the Ministry	The participation of the industrial producer leaves doubt on the objectivity of the information
ICE-UK www.bath.ac.uk/mech-eng/sert/embodied/	Building	Process enrgy and GHG	Compiles information internationally a number of sources such as government agencies or private companies Conducted within the University of Bath	Difficulty of control analysis and objectivity of the data
IBO Autriche www.ibo.at	Building	As a non-profit environmental impacts	Provides the primary data of global warming, acidification, energy, renewable, non-renewable, eutrophication, ecological index	Data based on the declaration of manufacturers with verification test

For ecological data, the necessity to provide values over the full life cycle as defined in ISO 14040 is accepted by all databases. It is also important to have the emission values for each phase of the life cycle.

2.2.3.1 Conclusion

The environmental field is documented accurately. Information is available and accessible. Multiple sources for environmental indices imply the need for a thorough qualitative analysis on the basis of established criteria.

2.3 Tools for Ecological Performance

2.3.1 Standards and Guides

Following the awareness of the impact of human activities on the environment and more generally on sustainable development, the need for decision support tools has ben revealed. This section summarizes the existing tools.

2.3.1.1 Standards

The ISO standards 14040—Environmental Management—Life Cycle Analysis (ISO 14040, 1997) appeared in 1997. They define the founding principles of the standards and tools whose objectives are impact measurements. This series of standards is based on two fundamental principles. The first is based on the fact that evaluation is meaningful only if it takes into account the entire life cycle. This principle, accepted by all stakeholders, has a recent and clear progress in the development of other standards and evaluation tools. The life cycle is defined as "cradle to grave" for all impacts related to the production, use and disposal. The second principle is the necessary inventory of impact elements based on the notion of delimitation of the system. This principle guarantees the exhaustivity of the impacts.

A wide methodology and framework, the ISO 15392:2008 standard—Sustainability in building construction—General principles—clarifies the application of the concept of level of sustainable development in buildings. It contains three main aspects, namely, the environment, the economy and society. Also at this level, with a methodological lens, ISO/TS 21929-1—Sustainable development in construction, provides a common frame of reference which will allow a better consideration of the performance of buildings with regard to sustainable development. It provides recommendations and guidelines for the development and selection of appropriate indicators of sustainable development for the building. Sustainability in building construction—Sustainable Development Indicators—ISO 21929-2 standard Part 2: Framework for the development of indicators for civil engineering is being developed and will be published in 2013.

In the scale of the building, the ISO 21931-1—Sustainability in building construction—Methodological framework for assessing the environmental performance of construction works—Part 1: Buildings, provides a set of criteria and reference principles. It proposes a framework for building owners and designers that will enable them to evaluate the performance of their projects.

EN 15643-1, 2: 2010—Contribution of construction to Sustainable Development—Evaluation of the contribution to sustainable development of buildings—presents general principles and requirements for the assessment of buildings in terms of environmental, social and economic performance. This evaluation aims to quantify the contribution to sustainable development by construction works. It clarifies again the definitions of the principle including the very interesting concept of functional equivalent. It is stated that « comparisons between the results of assessments of buildings or assembled systems (part of the work), when designing or whenever the results are used, should only be made with reference to the functional equivalent of building. This requires that the main functional requirements are described with the intended use and the relevant specific technical requirements. This description is used to determine the functional equivalence of different options and types of construction and forms the basis for a transparent and reasonable comparison. » For each phase, the conditions for an approach to sustainable development are described. EN 15643-3—Contribution of construction works to Sustainable Development—Evaluation of buildings—Part 3: Framework for the assessment of social performance is being validated. It will be published in March 2012.

ISO 21930 : 2007—Buildings and built structures—Sustainability in building construction—Environmental declaration of building products incorporates the principles of the standard NF P01-010 for construction products internationally. It establishes a clear set of indicators to quantify the production, implementation, use and demolition of building products. Prior to this and based on this standard, NF P01-010: 2004 Environmental quality of construction products—Environmental Statement and health of construction products (NF P01-010, 2004) has enabled the establishment environmental and health reporting sheets on the basis of the inventory of impacts, considered over the entire life cycle. This standard allows the establishment of true environmental and health hazard identity card of building products. The principles of business are specifically described in Appendix 3.

2.3.1.2 Benchmarks

There are numerous benchmarks in the building industry. Some of the major ones are HQE, LEED, BREEAM, CASBEE…. Detailed analysis of the main benchmarks is presented in Appendix 1. The summary of this analysis can show the shallowness of taking into account the economic dimension, probably due to the difficulty of its establishment. Furthermore, taking into account the lifespan is rare.

In France, the HQE certification has been analysed in detail (Hetzel 2009). The objective of benchmarks is to set a common set of rules to guide designers and builders. Eventually, this can result in a certification Benchmarks are also performance evaluation and diagnostic tools. It is only recently, in its 2008 version that the HQE Benchmark for environmental quality of buildings, encourages consideration of adaptability on short, medium and long terms up until 100 years. For other standards, when the latter is taken into account, it ranges from 30 to 90 years.

2.3.2   The Analytical Tools for Evaluating the Performance

The performance assessment tools are those involved in the category of tools that assist decision making. They are used at an advanced stage in design projects or at the diagnostic stage for existing projects. They quantify phenomena and thus identify and compare the performance of solutions. As there are numerous assessment tools, a comprehensive analysis was not possible. However, a detailed description for the most popular tools has been provided in Appendix 2.

In 2002, a state-of-the-art survey by the HQE2R group mentioned the extreme diversity of indicators using existing tools. The conclusion is that it is not feasible to have a selection of indicators which is common to the whole European community, and that a systematic adaption of the tools is therefore necessary.

In 2006, following careful analysis of existing tools, Ness B concluded that all the evaluation tools focus on the environment (Ness 2006). There are no tools addressing the social and economic aspect other than those dealing with costs. Indicators are not integrated and evaluations are conducted on local or national scales. Moreover, their analysis reveals an extreme simplification when taking into account the lifetimes of products or buildings evaluated. Furthermore, lifetimes considered are relatively short since the longest is 80 years, and none of these tools actually assess the impact of lifetime. Another observation emphasizes the scarce use of economic indicators.

Concerning the state of the art in tools that evaluate performance in terms of sustainable development, Haapio carefully describes their conditions (Haapio and Viitaniemi 2008). In summary, the researcher points out that the comparison of the tools is difficult or impossible. Tools mainly concern ecological evaluation. The overall assessment of the performance in terms of sustainable development seems unattainable.

Concerning tools assisting decision making for property management, only ASCOTT and APOGEE—PERIGEE models deal with overall costs and the economic aspect in detail. The decision-assisting tool BEES and repository BREEAM briefly evoke discounting. It seems that for the latter, no development is introduced as regards the causes and consequences of the choice of rate, which leaves the users to their own expertise. Other tools include no economic module.

The analysis of these tools suggests two conclusions. First, no integrated assessment tool exists. Evaluations are essentially ecological and except for the issue of costs, no tool addresses the social and economic aspects. Second, evaluations are conducted on national or local levels. At the building level, the assessment tools deal with thermal and environmental aspects.

2.3.3 The Models

In 2001, Voglander proposed the Eco-cost Value Ratio (EVR). This is a model which aims at the optimization of the design on the basis of efficiency (Vogtländer et al. 2001). The assessment of environmental costs and the fight against these costs as part of a response to a specific need provide, according to the author, the most effective approach in terms of sustainable development. This is an approach linking economic and environmental impacts. It has the distinction of introducing the concept of full costs by taking into account the costs of externalities. This workable model in all sectors therefore proposes to establish a ratio EVR = Ecocosts/Value. The Ecocost is the sum of the costs of toxic emissions, energy consumption resources, depreciation of equipment and human labor. Its value is that of the goods produced. This is a model of efficiency, with the object of taking into account and reducing environmental impacts. Lifeitme is indirectly integrated through the value of the goods.

Zhang has developed an evaluation model of the environmental performance of buildings, BEPAS (Zhang 2005). After listing and weighing the overall environmental impact of a building, the model makes an aggregate of its relative impact and supplies a final score. Again, the lifetime for the evaluation is fixed a priori—in this case 50 years.

Alwaer furthers the study of taking multiple criteria into account by looking at the distribution of key indices in an approach to evaluate the performance of sustainable development (Alwaer and Clements-Croome 2010). There is no reference to the inclusion of lifetime in the choice of appropriate indicators.

N. Banaitiene presents a multicriteria evaluation method in the complete life cycle of the building (Banaitiene et al. 2008). The method at no time refers to the potential impact the life of the building—neither to the selection criteria, nor to the consequences of the latter.

References

Alwaer H, Clements-Croome DJ (2010) Key performance indicators (KPIs) and priority setting in using the multiattribute approach for assessing sustainable intelligent buildings. Build Environ 45(4):799–807

Banaitiene N, Banaitis A, Kaklauskas A, Zavadskas EK (2008) Evaluating the life cycle of a building: a multivariant and multiple criteria approach. Omega 36(3):429–441

Bernstein D, Champetier JP, Hamayon L, Traisnel JP, Vidal T (2006) Traité de construction durable : Principes et Détails de construction. Le Moniteur Editions, 15 Déc 2006

Boothroyd P (1990) L'évaluation environnementale, un outil de développement durable équita-
ble. Dans développement durable et évaluation environnementale: perspectives de planifica-
tion d'un avenir commun Jacobs et Sadler, (dir.), pp 159–172

Boulanger J-M (2004) Les indicateurs de développement durable : un défi scientifique, un enjeu
démocratique. Institut pour un développement durable, Belgique1 Juillet 2004, Séminaire
développement durable et économie de l'environnement, Séminaire de l'IDRRI

Clift R (2004) Metrics for supply chain sustainability. In: Technological choices for sustainability. Springer, Heidelberg, pp 239–253

Curran MA (2006) Life cycle assessment: principles and practice

Déoux S, Déoux P (2004) Le guide de l'habitat sain. 2 éd. Medieco, 30 Sept 2004

Dufresne J-L, Salas Y Mélia D, Denvil S, Tyteca S, Arzel O, Bony S, Braconnot P et al (2006)
Simulation de l'évolution récente et future du climat par les modèles du CNRM et de l'IPSL
(Recent and futur climate change as simulated by the CNRM and IPSL models). http://hal.
archives-ouvertes.fr/hal-00423543/

Enghoff MB, Svensmark H (2008) The role of atmospheric ions in aerosol nucleation ? a review.
http://hal-insu.archives-ouvertes.fr/hal-00304109/. Accessed on 17 April 2008

Haapio A, Viitaniemi P (2008) A critical review of building environmental assessment tools.
Environ Impact Assess Rev 28(7):469–482

Habert G, Bouzidi Y, Chen C, Jullien A (2010a) Development of a depletion indicator for natural
resources used in concrete. Resour Conserv Recycl 54(6):364–376

Habert G, Bouzidi Y, Chen C, Jullien A (2010b) Development of a depletion indicator for natural
resources used in concrete. Resour Conserv Recycl 54(6):364–376

Hetzel J (2009) Indicatuers du développement durable dans la construction. Edition AFNOR

Jevons WS (1866) The coal question: an enquiry concerning the progress of the nation, and the
probable exhaustion of our coal-mines. Macmillan, London

Le Treut H et al (2008) Incertitudes sur les modèles climatiques. Géoscience. http://france.
elsevier.com/direct/CRAS2A/

Liébard A, De Herde A (2006) Traité d'architecture et d'urbanisme bioclimatiques : Concevoir,
édifier et aménager avec le développement durable. Le Moniteur Editions, Mars 30

Liu M, Li B, Yao R (2010) A generic model of exergy assessment for the environmental impact
of building lifecycle. Energy Build 42(9):1482–1490

Lorius C (2003) Effet de serre: les lacunes du savoir et de la perception: Greenhouse effect: gaps
of knowledge and perception. Comptes Rendus Geosci 335(6–7):545–549

Ness B (2006) Categorising tools for sustainability assessment.pdf. Ecological economics

Oberg M (2005) Integrated life cycle design-applicd to concrete multi-dwelling buildings

Osso A (1996) Sustainable building technical manuel—green building design, construction, and
operations. Public Technology Inc, USA, p 292

Peuportier B (2001a) Training for renovated energy efficient social housing—section 2 tools.
Intelligent energy europe programme, contract n° EIE/05/110/SI2.420021

Peuportier BLP (2001b) Life cycle assessment applied to the comparative evaluation of single
family houses in the french context. Energy Build 33(5):443–450

Pulselli RM, Simoncini E, Pulselli FM, Bastianoni S (2007) Emergy analysis of building manufacturing, maintenance and use: Em-building indices to evaluate housing sustainability.
Energy Build 39(5):620–628

Vogtländer JG, Hendriks PCF, Brezet PHC (2001) The EVR model for sustainability–A tool to
optimise product design and resolve strategic dilemmas. J Sustain Prod Des 1(2):103–116

WCED (1987) Our common future. Oxford University Press

Zhang Z (2005) BEPAS—a life cycle building environmental performance assessment model.
Build Environ

Chapter 3
Research Analysis

Abstract This chapter presents the analysis of the state of the art concerning the issue of lifespan of the building and its impact during the application of the concept of sustainable development in the building.

Keywords Lifespan • Building • Sustainable development

Studies of the environmental impacts related to the production phase of housing, which take into account the lifetime of the buildings, are scarce and lack thoroughness. We can explain and analyse the following work.

Thormark studied the energy impact associated with the choice of different products for townhouses over a period of 50 years (Thormark 2006). She showed how in the "low energy" buildings "energy method" has gained prominence. The amount of energy consumed during use goes from 85–90 to 40–60 % of the total energy consumed. The choice of products and their recycling potential become predominant in terms of performance. For the researcher, the total energy consumed in "low consumption" buildings may be higher than in those with high consumption. The researcher concludes on the designer's need to not only work on performance but also to choose products which are easily recyclable.

In 2007, Werner concluded on the superiority of performance-based products compared to other wood products, provided they are used appropriately (Werner and Richter 2007). These products are involved in the fight against greenhouse gas and produce less waste. However, the author warns the reader about impregnations which can cause poor performance. He particularly emphasizes the harmfulness of products containing adhesives, resins and varnishes. The lifetimes on which the simulations are performed range from 30 to 60 years.

Gustavsson and Sathre compared only solutions of wood and concrete for estimating CO_2 emissions (Gustavsson et al. 2006). They conclude on the real importance of wood-based products. Note that the lifetimes of the building for the study is fixed. It is 100 years.

M. Méquignon and H. Ait Haddou, *Lifetime Environmental Impact of Buildings*,   45
SpringerBriefs in Applied Sciences and Technology, DOI: 10.1007/978-3-319-06641-7_3,
© The Author(s) 2014

Peuportier conducted an ecological comparison of three detached houses over a period of 80 years while the differences in energy consumption during use and the other phases are not distinguished in the comparison (Peuportier 2001). The three buildings compared are neither equivalent in terms of energy consumption during use, not in terms of area.

Häkkinen studied the environmental impacts assessed on a complete cycle of a facade made of wood according to its treatment. The evaluation was carried out over a period of 100 years (Häkkinen et al. 1999).

In these studies, the lifetimes are introduced as a fixed datum. Whether it concerns energy studies, greenhouse gas emissions or wider environmental impact assessment, lifetimes are always less than 100 years and are fixed. It is only the performance of the solutions which is evaluated. The impact of lifetime has never really been thoroughly studied.

Regarding the evaluation of performance in terms of lifetime, there is very little work. In 2005, Mora questions the importance of life and impact on performance in terms of sustainable development of new materials for which we lack perspective. He referred to the current risks of presenting a degraded view of our generation to future generations in terms of art and culture (Mora 2007). Mora discusses briefly the impact of the lifetime of buildings on sustainable development. He thus observes, schematically, that by increasing "from 50 to 500 years the lifetime of buildings, environmental impacts are reduced by a factor of 10." The scarcity of recycled materials is also mentioned. Finally, the article distinguishes the duration of the work and the durability of materials. These issues are briefly discussed without providing thorough answer. Finally, he mentioned the possibility of ephemeral buildings constructed with permanent materials while buildings with more ephemeral materials may be permanent if the maintenance and repair have been prescribed. Again, the issues are not argued. This article discusses some tracks of interesting principles. However, it does not indicate any quantitative proof or even the scale of the impact of lifetime.

The works presenting the most in-depth analysis are those of Haapio and Viitaniemi (2008). These works show the impact of various technical solutions in the life cycle of a housing building depending on whether they are 60, 80, 100, 120 or 160 years old. The authors successively vary the life of various insulators, wall elements, windows and covers. Evaluations are performed for different impacts: energy source, waste, air pollution, water pollution, climate change, natural resources. The structure is maintained, only the various components are in comparison with different durations. Assessments of the energy "processed" in different technical solutions are carried out but the differences in thermal performance were not taken into account and neutralized. Differences in heating consumption are included in the results. The study focuses on the evolution of impacts during the development lifecycle. The study does not cover the impact of the life of the building in response to a functional need. Comparisons are made on simple change of life and not the answer to a need for a fixed duration. The impact of the life is taken into account only partially.

3.1 Conclusion

The necessity of taking into account the full life cycle of products or building regardless of the area evaluated is recognised.

Concerning the links and the impact of the life of the building or the components on the performance of sustainable development factors, there has been no study where attempts are made to measure its importance. The evaluation times are always short—around 50 years and sometimes up to 100 years. These times are chosen as fixed parameters. Undoubtedly, the impact of the value of the lifetime is not analysed or quantified.

The question we therefore ask is to know what is the impact of the lifespan of the Mora building on the environment.

References

Gustavsson L, Pingoud K, Sathre R (2006) Carbon dioxide balance of wood substitution: comparing concrete- and wood-framed buildings. Mitig Adapt Strat Global Change 11(3):667–691. doi:10.1007/s11027-006-7207-1

Haapio A, Viitaniemi P (2008) A critical review of building environmental assessment tools. Environ Impact Assess Rev 28(7):469–482

Häkkinen T et al. (1999) Environmental impact of coated exterior wooden cladding. VTT Technical Research Centre of Finland, Finland

Mora P (2007) Life cycle, sustainability and the transcendent quality of building materials. Build Environ 42(3):1329–1334

Peuportier B (2001) Training for renovated energy efficient social housing—Section 2 tools, intelligent energy—Europe programme, contract EIE/05/110/SI2.420021

Thormark C (2006) The effect of material choice on the total energy need and recycling potential of a building. Build Environ 41(8):1019–1026

Werner F, Richter K (2007) Wooden building products in comparative LCA. Int J Life Cycle Assess 12(71):470–479

Part II
Method and Application

This chapter defines the framework, the method, and the elements of the study. The first section defines the precise limits of the wall unit that we intend to study for its behaviour in terms of greenhouse gases through its lifetime. The second section fixes how the simulations are performed, stating the scales of evaluation and the various indicators whose evolution is observed. The data sources exploited are presented and analysed. The last section describes the method used to achieve our objectives.

Chapter 4
Method

Abstract The purpose of this chapter is to determine the questions addressed and the scales of evaluation. The indicators evaluated and the data sources will be stated. The technical solutions to be compared, again precisely defined, are identical in the responses they provide to a functional need.

Keywords Building • Tools • Sustainable development • Assessment

4.1 Wall Unit

If the behaviour of a wall unit is to be evaluated and analysed, it must be delimited and characterized. The first part considers the object as a system. It is described through its function, its structure, and the links it has with its environment. The following part defines the scales of evaluation.

4.1.1 Delimitation of the System

4.1.1.1 Definition of the System

The Functional System

The question posed implies that the function of the wall unit under study must be precisely defined. In order to carry out a study over a long period of time, we chose a housing structure. Housing fulfils the principal function of "provide a place to live" and the primary function of "shelter from outside stresses". Besides the fact that this utilization function has not so far been treated in depth for the question posed by our subject, the long duration of the function enables simulations to be envisaged. The dwelling considered is an individual house in order to avoid any difficulties in imputing the impacts of shared premises. Taking communal parts into consideration would make it necessary to introduce distribution keys

M. Méquignon and H. Ait Haddou, *Lifetime Environmental Impact of Buildings,*
SpringerBriefs in Applied Sciences and Technology, DOI: 10.1007/978-3-319-06641-7_4,
© The Author(s) 2014

Fig. 4.1 Functional
analysis—components of the
building housing

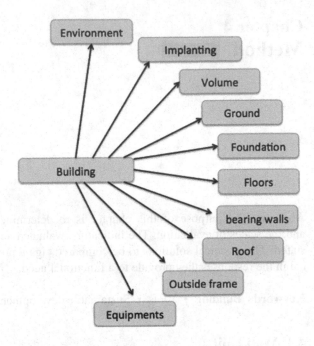

that would lead to complications and thus to uncertainty in the results and conclusions while, in fact, not contributing anything to the demonstration.

The Structural System

Based on subassemblies that compose the house analysed as a system, we can draw up the following simplified diagram (Figs. 4.1, 4.2).

Functional analysis—Element of the house structure. Our aim is to assess the impact that the lifetime of the dwelling and the technical solutions have on sustainable development. The elements marked with an asterisk in the system diagram above are second-level responses of the top-down functional analysis (FAST and standard NF X50-150). These solutions are always responses to a precise functional specification. The system approach favours the performance aspect of the responses to the functions imposed by the functional requirement.

4.1.1.2 Other Characteristics of \the System

Choice and Delimitation of the System in Time

To reply to our question, we define the lifetime of the function that the technical solutions must satisfy. The lifetime of the function "provide a place to live" is a hypothesis. Observation of existing dwellings leads us to propose a function lifetime of 300 years. This may seem a long time, but the choice reflects both the

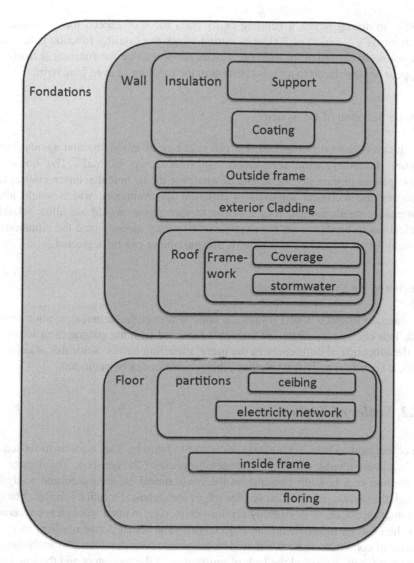

Fig. 4.2 Building dependency diagram for lifespan

problem and the large number of century-old buildings in our cities. The needs can be considered as temporally unlimited and there seems to be no reason for human physiological needs or our need for security and belonging, in Maslow's sense, to change whatever the time scale considered. Qualitatively, the needs met by the function may change but old houses can be adapted to changes in the original needs or to meet new ones (Philippe 1997; Latham 2001).

They still fulfil their service function, including in the sense of standard ISO 15686-1: 2005 Buildings and constructed assets—Service life planning—Part 1: General principles and framework, which defines the service life as the "Period after installation during which a building or its parts meet or exceed the performance requirements". A period of 300 years during which the housing function is fulfilled does not seem excessive. In addition, it does not seem that the function is likely to disappear or be reduced to any great extent in the medium or even long term.

Climatic Situation of the System

The situation chosen is standard, i.e. the intensity of environmental agents: temperature, humidity, wind or corrosive substances, is "normal". The house is always placed in the same environment whatever the technical solution envisaged, as putting the technical solutions in different environments, which would alter maintenance needs and the results of the service phase, would not allow reliable conclusions to be drawn on the impacts of lifetimes alone. Since the situation is identical for all the technical solutions, this parameter can be neglected.

Situation of the System in Space

The situation in space is also always the same whatever the technical solution envisaged. This choice neutralizes the impacts connected with the geographical location, e.g. the influence of differences in the users' travelling habits. Since the situation is identical for all the technical solutions, this parameter can be neglected.

4.1.2 Scales of Evaluation

The object studied here is a building dedicated to housing. This leads us to fix scales for the environmental, economic and social aspects of the question. The impacts of the lifetime of a housing building on the environment are evaluated and analysed through the consequences of the service life of the technical solutions chosen. For the economic aspect, the scale of study is micro-economic; in other words it is concerned with the costs associated with the object itself. The social domain suffers from an absence of consensus on how it should be taken into consideration. The analysis performed in Chap. 1 showed the lack of uniformity of the indicators and the low level of maturity of the social and cultural approach, particularly at building scale. Since there are no uniform social and cultural indicators like those available for environmental or economic aspects, these questions will be approached through a more qualitative inventory and analysis of the impacts the lifetime of the object has on society.

4.1.3 Summary

The object studied is a building fulfilling the functions of an individual house for a duration of 300 years. The geographical situation is always the same, whatever the technical solution proposed. The stresses of the local environment are taken to be "normal". The house is the response to a set of needs corresponding to the programme. The response is made up of a set of elements meeting functional needs resulting from a system approach complying with the FAST technique and standard X50-150.

4.2 Choice of Indicators and Means Employed

Once the object and the scales have been defined, it is necessary to state the principles, the indicators developed and the data sources used for the evaluations.

4.2.1 General Principles of Evaluation

To carry out our evaluations, three principals were selected: the service provided was the same whatever the chosen solution, the complete lifecycle was considered, and the dependencies between certain elements of the building were taken into account.

4.2.1.1 Comparisons "Ceteris Paribus"

The objective being to bring out the impact of the lifetime on durable development, the solutions compared must meet the same service specifications. Comparisons must be made ceteris paribus as far as the response to the service function is concerned. For example, the thermal capacity of the various technical solutions must be identical so as to neutralize the impact of power consumption for heating. The service provided during occupation must be strictly the same. For this reason, the specifications of the objects under study have been published?

4.2.1.2 Life Cycle Evaluation

Evaluations of the indicators must take account of (1) the impacts of the entire life cycle of the products in the sense of standard ISO 14040 and (2) the number of cycles required to meet the need, with respect to the service lives of the products, for the duration fixed for the service function of the object, i.e. 300 years.

4.2.1.3 Dependencies Among Building Elements

Considering the links among the various elements that compose the building and to take the duration of their service lives into account, we use the dependency diagram below. This assumes that the inclusion of one element in another implies that the end of the lifetime of the latter induces the end of lifetime of the former. By convention, we state that the end of the life of the hierarchically superior element entails the end of life of the hierarchically inferior element. The converse is not true; the end of life of the hierarchically inferior element does not entail the end of life of the dominant element.

For example, the end of life of an outside wall, a hierarchically superior element, entails the end of life of the insulation, the hierarchically inferior element, whereas the end of life of the insulation does not entail the end of life of the wall. The end of life of the insulation of the wall implies the end of life of the included element, i.e. the support for the inside wall covering, such as plasterboard.

When a dependency exists, for the sake of simplification, it is assumed that the service life of the "inferior" element is a submultiple of the lifetime of the "superior" element.

Note: although they are hierarchically inferior elements, load-bearing walls are considered to be "associated" with the foundations as the end of life of the foundations implies the end of life of the load-bearing elements. The converse is taken to be true. It seems unlikely that a building would be taken down only to be rebuilt with an identical plan.

4.2.2 Choice of Indicators

The impact of the service life of the conceivable technical solutions techniques on the environmental factor depends on a large number of indicators. To measure how much impact these indicators have, a sensitivity analysis is necessary. This analysis will allow the number of indicators to be considerably refined without losing the accuracy or pertinence of the results. Factor analysis and principal component analysis are statistical tools base on mathematical rules that are capable of significantly reducing the degree of freedom of a system of indicators.

In the present work, we chose principal component analysis (PCA) to perform the sensitivity study.

Once the indicator has been stated, the evaluation of the environmental impact covers the entire life cycle of the solutions, together with the consequences of the service life, which induces the obligation to rebuild during the lifespan of the function, i.e. 300 years.

4.2.3 Availability and Choice of Data Sources

Several databases provide information on the environmental impacts of building materials and products. There is only one available database regarding costs, and the social approach does not supply any structured, uniform data. The aim of this section is to analyse the available data, choose the data to be exploited and justify these choices.

Table 4.1 Example summary of EPD multi brick cells

N°	Environmental impact	Indicator value per m² of projected area and per annuity	Indicator value per m² of projected area for the entire life
1	Consumption of energy resources (MJ)		
	Total primary energy	5.27	527
	Renewable energy	3.64	364
	Non-renewable energy	1.64	164
2	Resource depletion (ADP) (kg antimony (Sb) equivalent)	0.000361	0.0361
3	Total water consumption (litres)	0.386	
4	Solid waste (kg)		
	Waste recycled	0.522	52.2
	Waste disposed of	0	0
	Hazardous wastes	8.84×10^{-5}	0.00884
	Non-hazardous waste	0.216	21.6
	Inert waste	0.00195	0.195
	Radioactive waste	2.01×10^{-5}	0.00201
5	Climate change (kg eq CO_2)	−0.0191	−19.1
6	Atmospheric acidification (kg equivalent SO_2)	0.000458	0.0458
7	Air pollution (m³)	9.53	953
8	Water pollution (m³)	0.0187	1.87
9	Destruction of stratospheric ozone layer (kg CFC)	–	–
10	Photochemical ozone creation (kg ethylene equivalent)	9.67×10^{-5}	0.00967

4.2.4 Data Sources

4.2.4.1 Environmental Data and Critical Analysis

Analyses of the impact of lifetimes must be founded on simulations, which require data. The difficulty lies in evaluating the quality of the information available from the various sources.

We listed and analysed the available databases in Chap. 1. The analysis brings out two principal bases that are suitable for carrying out our simulations: the INIES and ECOINVENT-KBOB databases. The indicator evaluations are performed over the complete life cycle, as defined in ISO 14040, in both bases.

The French INIES database publishes Environmental and Health Declaration Forms (FDES) established according to standard NF P01-010. It provides precise information on the products.

The full report sums up the performances in the form of an environmental and health identity card, as shown in the example in Table 4.1:

Fig. 4.3 Extract-based ECOINVENT KBOB

The information is clear, free of charge and available at the website http://www.inies.fr.

The ECOINVENT-KBOB base presents general results per type of product. It thus gives information that is less precise but more complete in the sense that it introduces certain equipment items, e.g. for heating. The information is presented by means of a spreadsheet, an extract of which is shown below.

The possibility of choosing products from the INIES base seems to provide greater accuracy. The information items are numerous and can enable the method to be applied to other indicators. The information is easily accessible and free. The protocol used to establish the FDESs is precise. The fact that the database is supervised by the ministry and state or semi-state bodies, the evaluations performed on the products, certain of which are carried out by independent bodies, and the control system set up all make the quality of the data more credible.

The understandable desire of firms to influence the information in order to make the presentation of the impact results more "attractive" nevertheless obliges us to analyse the data with a critical eye. We compared the product values found in the INIES base with the "typical" values of the ECOINVENT-KBOB base. The ECOINVENT-KBOB base is established by the Office Fédéral des Constructions et de la Logistique (OFCL), a Swiss state-run organization (Fig. 4.3).

A comparison of data concerning GHG emissions from the two bases, established for selected products used in buildings, is given in the table below (Tables 4.2, 4.3).

The comparison shows differences on certain wood-based products. This comes from a quasi-opposition in the methodology for taking certain aspects into account, which we will come back to in the next section. The differences for other products hardly exceed 20 %. These differences may seem large but, if certain factors are taken into account, some of them become relatively small. Firstly, the ECOINVENT-KBOB base evaluates the impact of typical products and not specific ones. The data are thus averages. In addition, the INIES base considers the product in use. In other words, in these examples, the connecting joints have been taken into account. The basic ECOINVENT-KBOB unit is the mass of the typical product. However different products having the same service function may not

Table 4.2 Comparison of indicator values GHG (kg CO$_2$ eq) and bases INIES KBOB-ECOINVENT

	INIES	KBOB	Relative difference (%)
Glass wool	30.90	35.28	12.41
Mineral coating	5.13	5.41	5.09
Multi cell brick	92	74	20.04
Full solid clay bricks	85	57	33.22
Concrete block	16	24	33.77
Shuttered concrete	65	58	11.53
Aerated concrete	40	41	2.10
Wood frame	−10.76	3.89	136.16
Stone	51	42	18.55

Table 4.3 Differences on 300 years of foundation and INIES ECOINVENT-KBOB

	INIES	KBOB	Relative difference (%)
Glass wool	30.90	35.28	12.41
Mineral coating	5.13	5.41	5.09
Multi cell brick	279	336	16.88
Full solid clay bricks	170	161	5.59
Concrete block	163	131	20.07
Shuttered concrete	173	149	13.72
Aerated concrete	206	194	6.00
Wood frame	−10.76	3.89	136.16
Stone	54	48	11.23

have the same density. Differences may also arise when the energy used is taken into account. The type of energy and the way it is produced can vary from one country to another, thus changing the results for emissions.

The proposed method is independent of the source of the data used. Nevertheless, it is clear that the results obtained are dependent on the data and can diverge for different sources.

4.2.5 Contradictory Data in Environmental Databases

The comparison of the data reveals very different values for wood-based products. There is a marked divergence concerning the value of the index for the greenhouse gas impact, in kg CO$_2$ equivalent, of the products derived from wood. For example, the ECOINVENT-KBOB base gives positive values of greenhouse gas emissions for these products. In contrast, the (Austrian) ECOSOFT and (French) INIES databases show negative values. Although the previous subsection tends to justify the choice of the INIES base, the "climate change" indices for wood products make it indispensable to perform a specific analysis and adopt a clear standpoint.

4.2.5.1 Specificities and Difficulties Concerning GHG

The quantities of CO_2 stored result from the growth cycle and the mechanism of photosynthesis, which is, in itself, highly complex. On the basis of a diurnal, seasonal mechanism, the balances, delays and durations differ according to the species, the soil, the environment, etc. Nevertheless, it is certain that, during growth, the CO_2 balance is negative. In other words, in this phase, a tree stores more CO_2 than it releases. When growth is just beginning, CO_2 storage is fairly low, but increases as the tree comes into full growth. When maturity is reached (the forest is at its climax) the absorption/release budget is relatively balanced and the storage function is smaller.

The end of this cycle poses two questions. The first concerns the replanting of areas where trees have been felled. The French forestry authority (ONF) guarantees that such replanting takes place, but what happens in the private sector? This step is important if we want to be able to count on carbon storage close to that of fully grown trees in a reasonably near future. The second concerns the end-of-lifetime of the product. In the absence of information, the standard NF P01-010 considers that the product is scrapped. In case of incineration, how should the emissions be accounted for? What precautions are taken concerning any treatment the wood has received and any glue or finishes that have been applied?

Drawing up a budget involves taking account of the emissions in connection with the energy consumed for forest maintenance, and the felling, transport and implementation of the wood.

Another difficulty arises when we consider the value of waste material from forest maintenance and the exploitation of the wood for production. How is monetary value obtained for these waste products? How are they used? Are they transformed and thus stored or are they burnt?

Thus, precautions must be taken when evaluating the greenhouse gas emissions of wood products over a complete life cycle (Vial and Cornillier 2009). These precautions concern:

- the modelling of the forestry stage, which should the allow the specificities of the plant biology and the relationship between the various compartments, water, air, soil and vegetation, to be represented;
- the modelling of fluxes connected with the biomass carbon in life cycle inventories from forestry to end-of-life, and how impact indicators are taken into account in the calculations, in particular the climate change indicator. Their absence can induce effects that are counterproductive in the fight against GHG emissions;
- the modelling of temporary carbon storage during the wood utilization phase and inclusion of the benefit of this sequestration in the evaluation of the impact on the contribution to climate change of the system under study;
- the posting of energy, matter and procedures, definitions and choice of energy indicators;
- the choice of allocation among the products studied, the co-products and the by-products coming from the transformation of wood.

To finish, CO_2 storage is a proportional function of the volume of wood used but conserved. It is constant at identical wood volume and varies according to the variation of the volume of wood used.

4.2.5.2 Result of the Approach Omitting the Storage Function

Based on the complete life cycle, the position of some analysts on how to take bio-mass fluxes into account is to consider the overall ratio of absorption to emission as neutral. The carbon fixed is released during the life cycle and particularly at the end of life, either by incineration or by biochemical breakdown. The resulting GHG budget is small but the value of the indicator is positive.

4.2.5.3 Arguments for a GHG Impact Indicator with a Negative Value

The methodological choice of not taking account of fluxes connected with bio-mass carbon makes it impossible to evaluate the negative or positive impact of these fluxes. It sets a forest that is slashed and burnt on the same level as a forest that is sustainably managed. Conversely, taking biomass carbon into account could enable the distinction to be made between a wood product coming from forests managed in a sustainable manner and one resulting from deforestation, for which no claim to CO_2 storage could be made. Rabl thus concludes that failure to take account of emission and absorption connected with biomass carbon can lead to conclusions that are counterproductive in the combat against climate change (Rabl et al. 2007).

"ISO 14040, which gives the principles and framework of the life cycle analy-sis, does not state anywhere that the fluxes connected with biomass carbon should not be taken into consideration. With them, the climate change indicator can turn out to be negative, expressing a beneficial effect against climate change if the bio-mass carbon is not released" (Vial and Cornillier 2009). For these authors, this is all the more true as the slow breakdown of wood in a dump over 100 years, the time recommended by the GIEC for a working scale, allows us to consider that the objective is reached.

4.2.5.4 Difficulty for Our Method

Considering our method as illustrated by Chart 4.1 below, the negative-value indicator suffers from the following paradox. Since it is established for a com-plete life cycle, the shorter the service life of the product, the larger the num-ber units consumed to satisfy a functional need and the greater the benefit for the planet due to the effect presented. Thus the use of this index results in an apparent absurdity.

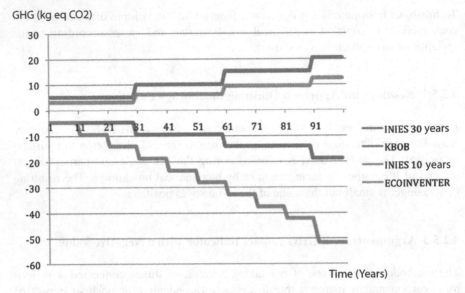

Chart 4.1 Cumulative GHG for "wood" solutions depending on the data source and the product service life

4.2.5.5 Position Adopted

For products having components made from wood materials, whenever the index leads to difficulties of interpretation as described in the previous subsection, we present the results connected with the INIES and ECOINVENT-KBOB data bases side by side.

4.2.5.6 Summary

The available data are sufficiently numerous and accurate to provide an answer to the question posed by our subject. The protocols for gathering the data are, themselves, available and precise. As far as environmental indices are concerned, we need to remain vigilant about the declarations made by manufacturers. Nevertheless, the relative consistency of the information in the INIES and ECOINVENT-KBOB bases enables us to feel fairly serene about using INIES data for our demonstration. For the GHG indicators for "wood" products, given the contradiction pointed out above, we shall keep the data from both the INIES and ECOINVENT-KBOB sources. Concerning costs, we take the only database serving as a reference for professionals: Batiprix. It is accurate and detailed, and so permits our question to be answered. The social approach does not use a database as such and a specific analysis method will be developed for this aspect.

4.3 Presentation of the Method

To achieve our objective and be able to study building performance in terms of sustainable development with respect to the lifespan, this section explains the strategy chosen, then the processing of the impacts in the occupation phase. The final part of the section will deal with the development of the method.

4.3.1 Analysis of Performance of the Wall Unit and the House

The house is considered as a system here, i.e. a set of elements that have relationships with one another and with the surroundings. The system always has the same geographical situation whatever the technical solutions chosen. This principle allows the impact factors that are not connected with the lifespan to be neutralized and neglected. For example, the energy requirements for the use of the house are the same for all the technical solutions and we do not evaluate the impact of the service lives of the technical solutions according to the geographical situation.

The first step of our approach is to establish and carry out the behaviour evaluation for an element of the house in terms of sustainable development over a long lifespan of 300 years.

There were several reasons for this choice:

- This element is one of the most durable of the whole building and thus responds more easily to the long time scale chosen in the hypothesis.
- It includes the important question of thermal behaviour. On this point, the choice of the element enables constraints to be fixed for reducing the impact during occupation and allows them to be solved with good accuracy.
- It allows a precise methodological protocol to be designed to answer the question.

The choice of this element is validated by the results. The basic element of the "vertical walls" component, presented in result 18, is validated as the second most emissive component.

However, although the behaviour of the unit allows comparisons and analyses to be made, it does not provide:

- the scale magnitudes through comparisons;
- the possibility of checking similarities in the responses of solutions during use;
- the information needed for making strategy and policy decisions concerning the question of the impact of the lifespan of buildings.

The values obtained when evaluating the behaviour of the wall unit have no common standard of measurement since a scale of the results cannot be produced as they stand. After having studied the possibilities and limits of using the method for other components of the house, we will extend the method developed for the wall unit to the whole of a typical house. For the parts of the house other than the walls, we do not propose several technical solutions; the solutions selected are the standard ones. Multiplying the factors to be modified, and that have an impact on the results, would only make the analyses and conclusions ambiguous. When the lifetime of the walls is introduced at whole-house scale, the impact function of the wall unit can be extended to the whole building by homothety, with the addition of a constant function for the impact of the other components. In the comparison exercise, the approach introducing the other components thus reduces the relative differences between solutions as they are "mixed in" with the other components having fixed impact.

4.3.2 Taking Impact of Utilization into Account

The accepted necessity to make evaluations over the complete life cycle implies that all stages of product life cycles must be considered: production, implementation, utilization, maintenance and demolition. This principle is shown schematically in (a) of figure.

4.3.2.1 Taking Normal Maintenance into Account

As the technical solutions have the same outside and inside finishes (except for the "stone" solution), for the sake of simplification, we assume that normal maintenance is the same for all the solutions. This identical maintenance does not alter (apart from a constant) the results of the impact evaluations and is thus not evaluated.

4.3.2.2 Taking Maintenance into Account

The chosen geographical situation does not lead to any particular stresses. The operations are assumed to respect the manufacturers' recommendations and standard practice. How the element under study, i.e. the wall unit, is kept operational is considered in the framework of the service lifetimes of the products that compose it.

4.3.2.3 Repair Work

For the life spans of the technical solutions taken in our hypothesis, it is assumed that there are no incidents that lead to repairs. This hypothesis forms part of the restrictive conditions of the study.

4.3.2.4 Neutralization of Utilization Impacts

The aim is to study the impact of the service lives of the technical solutions on performance in terms of sustainable development. The service lives and their impacts, which differ according to the solutions, make it obligatory to compare systems that satisfy strictly the same needs and in the same conditions. The technical solutions proposed must be subject to the "ceteris paribus" clause. This equivalence is indispensable if the impacts of the service lives of the different solutions selected are to be compared. An absence of equivalence in the response to needs would make the comparison of solutions inaccurate and the conclusions uncertain. This strict equivalence of response to utilization must therefore entail the strictest possible equivalence of environmental and economic impacts for the housing function during its use. For example, the winter comfort situations must be as similar as possible whatever the technical solutions chosen. This equivalence of behaviour during utilization should be verified with the tools available. It is the design characteristics of the technical solutions that enable the impacts of the utilization phase of the "housing" function to be made equal. It is thus the incidences of the design that count in the simulations. For the same utilization, the differences in behaviour that cannot be reduced through design and that generate differences during use have to be incorporated in the simulation results, unless they are negligible. The objective is represented in (b) of the diagram below.

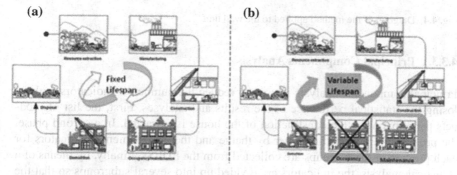

Mapping phenomena taken into account in the analysis of classical life cycles (a) (after www.fs. fed.us) and compared our approach presented (b)

4.3.3 Development of the Method

The approach comprises three stages. The first solves the problem arising from the large number of environmental indicators. The second consists of defining a precise element composing the "housing" system, viz. the wall unit. This stage develops the methods, clearly defines the element under study and the limits, and finally presents the behaviour of the various solutions over time according to the hypotheses made for their lifetimes. The third stage extends the previously defined method to the system constituted by a complete typical house.

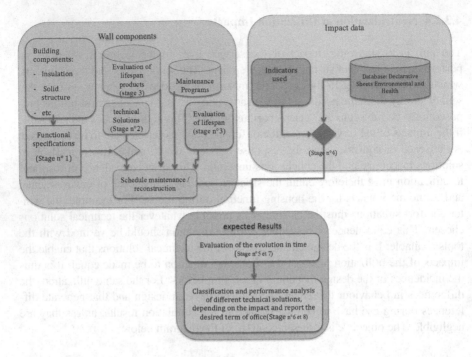

Fig. 4.4 Diagram of the method applied to the wall unit

4.3.3.1 Principal Components Analysis

Principal components analysis is used to reduce the number of indicators without losing the statistical pertinence of the results and analyses. First, the list of products participating in the construction of the house is extracted. In a second phase, the products are grouped together by theme and the environmental indicators for each product of these groups are collected from the FDESs. Finally, by means of a statistical analysis, the indicators are divided up into several subgroups so that the most representative can be selected. The products whose indicators are taken into account are the products composing a building intended for housing and those that are part of the various technical solutions envisaged.

4.3.3.2 Definition and Study of the Unit of Outside Load-Bearing Wall

The objective in studying this element was to be able to simulate the behaviour of the environmental indicators and the costs according to the hypotheses chosen for the service lifetimes of the technical solutions.

The method is represented schematically in Fig. 4.4 and explained in greater detail below.

Table 4.4 Technical
solutions to meet the
specifications

Technical solutions used for the wall
Concrete block (20 cm)
Wooden structure (23 cm)
Multi cell brick (30 cm)
Stone (24 cm)
Aerated concrete (20 cm)
Brique terre cuite pleine 200
Béton banché
Mud brick 200 mm
Full solid clay brick (20 cm)

Note: Considering Fig. 4.2, to simplify matters, the link between the wall and the outside woodwork, the floors and the roof are ignored, as are the corresponding impacts. This simplification is permissible because the lifetimes of the roof and joinery will always be taken to be shorter than or equal to those of the wall.

The goal of the study is thus only to evaluate the performance of the wall itself, by considering how its performance levels are maintained.

Following the progression of Fig. 4.4, we now present the various phases of our approach.

Phase 1: Analysis of need, functional specifications, expected characteristics

This stage guarantees the equivalence of the solutions proposed in response to the utilization function. The specifications ensure that the divergence in the functional responses is reduced. For example, it is necessary to neutralize the differences in energy consumption for heating by respecting the same insulation coefficient and to put forward solutions that provide the same decorative effect and comfort.

Starting from the methods initiated by the standard NF X 50-151 and the APTE method, recognized and adopted in industry for the design phase, we state the utilization functions and draw up an overall functional specification to be followed in the design phase and for the choice of technical solutions. The method consists of situating the element under study in its system by describing the links it has with the other elements and its interactions with the elements of its outside environment. These links are expressed as functions for which the criteria and levels of criteria are defined. All this enables the full specifications to be drawn up.

Phase 2: Development and presentation of the various technical solutions

This stage consists of proposing various technical solutions that fulfil the utilization functions while respecting the specifications of the previous phase. The different technical solutions selected must enable impacts to be compared in terms of sustainable development over the duration fixed for the defined need. The technical solutions chosen must be readily available. The solutions that meet the specifications include the structural solutions presented in the table below (Table 4.4).

The criteria used to make these choices were: modern-day practice, practices that were current in the past, and solutions that are starting to impose themselves.

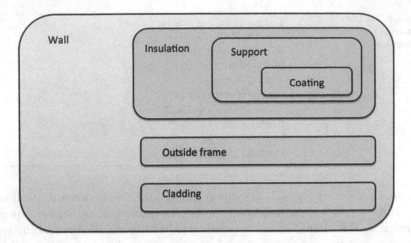

Fig. 4.5 Dependency of lifespans of the wall

The technical solutions proposed should lead to reflection on the nature of the impacts of these solutions on other elements of the system and on the upkeep that a solution imposes to maintain its performance level.

Phase 3: Evaluation of service lives of the proposed solutions

The service life evaluations can be the results of experimental, reliability or statistical approaches or based on "expert opinion" (Talon 2006). It is this "expert opinion" method that we shall use to choose service life intervals. As a first approach, this method allows a simulation to give approximate results based on fuzzy logic (Bouchon-Meunier and Marsala 2003). To define the service lives of the elements composing our wall unit and the frequency at which they should be replaced, we return to the part of Fig. 4.2 corresponding to the wall, in the diagram below (Fig. 4.5).

For the sake of simplification, the relationship with the outside woodwork is neglected. The inside wall covering, with a short lifespan, is also excluded.

Phase 4: Collecting the values of the chosen indicators

The various criteria of the study having been defined in the preceding stages; the work in this stage consists of bringing together the data that enable the chosen indices to be evaluated and presenting the results and their evolution over the chosen duration. The databases are INIES and ECOINVENT-KBOB for the environmental aspect.

Phase 5: Environmental and economic simulations

A variety of results can be obtained from the simulation of how the impacts evolve, on the basis of the chosen hypotheses. The principal processing applied to the data uses a spreadsheet to accumulate the impacts over the duration of the function.

The expected results are the following:

- evolution of cumulative greenhouse gas emissions over the time the need is defined to last, i.e. 300 years;
- comparison of the results with the official indicator values;
- evolution of cumulative costs;
- links established between costs and GHGs.

Note: Although a social approach to the impact of the wall unit lifespans would be interesting, this aspect will only be considered at whole-building scale.

Phase 6: Analyses of results

This phase consists of comparing and analysing the various results.

Phase 7: Variable lifespans for the technical solutions

The service life of a product is difficult to determine. We call columns 2 and 3 of Table 4.2 into question and so, during this phase, the lifespans of the different solutions are considered as variable. Evaluation is performed by fixing the lifetimes of the products a priori. This step assesses the impact of lifespan of the solution itself on its own results, but it also allows the solutions to be compared with one another.

The results expected are:

- Evolution of impacts according to the lifespans of the solutions;
- Evolution of costs according to lifespan.

Phase 8: Analyses of results

The evolution of the indicators according to the lifespan can be precisely identified. Moreover, this phase establishes correspondences among the various technical solutions on the basis of their having the same impact.

4.4 Conclusion

In this chapter, the questions addressed and the scales of evaluation have been precisely determined. The indicators evaluated and the data sources have been stated. The technical solutions to be compared, again precisely defined, are identical in the responses they provide to a functional need. The neutralization of the impacts during utilization, which has been verified, should allow comparisons to be made ceteris paribus, among the technical solutions as far as the impacts of production, maintenance and end-of-life are concerned. The expected results should, in particular, permit an evaluation of the consequences the lifetime of the building has on the chosen indicators, viz. greenhouse gases and the economic factor. Finally, an inventory of the social consequences of the building's lifetime can be made.

References

Bouchon-Meunier B, Marsala C (2003) Logique floue, principes, aide à la décision. Eds. Lavoisier Paris

Latham D (2001) Creative re-use of buildings, vol 1. Donhead, Shaftesbury

Philippe S (1997) Architectures transformées: Réhabilitations et reconversions à Paris. Paris: pavillon de l'Arsenal

Rabl A (2007) Interpretation of air pollution mortality: number of deaths or years of life lost? J Air Waste Manag Assoc 53(1):41–50

Talon A (2006) « Evaluation des scénarii de dégradation des produits de construction » Thèse Génie Civil. Clermont-Ferrand : Centre Scientifique et Technique du Bâtiment – service Matériaux et Laboratoire d'Etudes et de Recherches en MEcaniques des Structures » , 2006, 240 p

Vial E, Cornillier C (2009) Accounting for temporary biomass carbon storage in environmental labelling. In: Proceedings of the international conference on carbon storage in wood products, Brussels

Chapter 5
Application

Abstract The purpose of this chapter is to validate and implement the method described in the previous chapter by using selected data. The procedure used here is the reduction of data into principal components.

Keywords Indicator value • Data selected • Principal components analysis PCA

5.1 Data Selected

The procedure used here is the reduction of data into principal components. In the first phase, for each major element of a building (floors, load-bearing walls, insulation, frame, etc.), we select several products that can meet the needs. These products and their FDESs are presented in Table 5.1.

The wood frame, fired solid clay and unfired clay solutions do not have specific FDES. A specific evaluation was performed for them on the basis of similar products. In the table below, we bring together data provided in the FDESs (Table 5.2).

5.2 Indicator Values

The wood-base products, such as the frame, have a performance profile that is very different from those of the other products. For example, the way greenhouse gas emissions are calculated for these products gives them very specific behaviour with negative values. These products were excluded from the principal components analysis.

When indicator values for wood-based products were compared with the values of the indicators of the other products, the following correlations were found (Table 5.3):

M. Méquignon and H. Ait Haddou, *Lifetime Environmental Impact of Buildings*,   71
SpringerBriefs in Applied Sciences and Technology, DOI: 10.1007/978-3-319-06641-7_5,
© The Author(s) 2014

Table 5.1 Inventory EPD

Products	EPD reference	Vérification	Date
Concrete block	Wall small hollow concrete elements	Vérified	September 2006
Aerated concrete	Masonry wall blocks Aerated concrete	BIO intelligence service	November 2007
Mineral coating	Mortier d'enduit minéral	ECOBILAN	January 2007
Glass wool	ISOCONFORT 35 220 mm thick glass wool	ECOBILAN + Vérified	January 2006
Shuttered concrete	C25/30 concrete wall XF1 CEM II with complex thermo-acoustic insulation	ECOBILAN	October 2007
Multi cell brick	Monomur 30 cm	Vérified	October 2009
Hourdis floor	Hourdis floor	No verified	
Outside frame	Window in scots pine	Vérified	March 2008
Roof covering	Masonry clay element	Vérified	October 2011
Stone	Masonry element Pierre de Noyant	No verified	June 2010

This matrix reveals very strong correlation between:

- Non-renewable energy/Primary energy;
- Process primary energy/Primary energy;
- Resource depletion/Primary energy;
- Resource depletion/Non-renewable energy;
- Resource depletion/Process primary energy;
- Resource depletion/GHG; and
- Between hazardous waste and the "Renewable energy" indicator.

Table 5.2 Values of the indicators contained in the EPD

	Total primary energy (MJ)	Renewable Energy (MJ)	Non-renewal energy (MJ)	Primary energy process (MJ)	GHG (kg éq CO2)	Hazardous waste (kg)	Non-hazardous waste (kg)	Inert waste (kg)	Radioactive waste (kg)	Air pollution (m³)	Water pollution (m³)	Acidification (kg éq SO2)	Resource depletion (sb)	Water (l)	Reused waste (kg)	Photochemical ozone formation (kg ethylene eq)
Concrete block	174	15	158	166	16	1.25E−02	0.867	236	1.49E−03	1673	7.83	7.16E−02	5.74E−02	83	0.617	6.65E−03
Aerated concrete	485	29	457	479	40	4.00E−02	3	91	1.07E−03	2234	56	1.00E−01	1.57E−01	267	9	2.67E−03
Mineral coating	61.5	8.8	53	45.7	5.15	4.10E−03	24.15	0.21	4.75E−04	290	3.47	1.90E−02	1.89E−02	23.2	0.79	2.42E−03
Glass wool	128	24.3	104.5	0	4.16	3.78E−03	4.645	0.004	7.50E−04	670	1.3	5.80E−02	3.00E−02	33.7	0.3785	4.21E−03
Shuttered concrete	878	25.6	853	716	65.1	7.39E−02	25	242	4.78E−03	639	2.96	2.83E−01	3.33E−01	335	168	2.29E−03
Multi cell brick	824	115	709	787	41.3	2.61E−02	0.135	262	3.03E−03	2884	11.2	9.46E−02	2.37E−01	67.7	2.64	2.67E−03
Hourdis floor	456	11.1	445	333	31	4.70E−02	3.04	115	2.12E−04	2030	11.7	1.50E−01	1.23E−01	129	64.1	1.61E−02
Outside frame	1191	363	831	878	23.49	5.43E−01	11.76	16.98	3.99E−03	3720	72	2.31E−01	2.48E−01	187.8	23.43	0.00E+00
Roof covering	185	0.248	184	178.4	9.3	4.79E−03	0.912	46	7.60E−04	601	21.7	3.46E−02	6.74E−02	21.7	0.512	3.70E−03
Stone	328	69.8	258	259	24	6.90E−02	0.244	38	2.42E−03	2240	3.46	1.46E−01	9.20E−02	60.4	526	0.00E+00

Table 5.3 Correlation matrix

	Total primary energy	Renewable energy	Non-renewal energy	Primary energy process	GHG	Reused waste	Hazardous waste	Non-hazardous waste	Inert waste	Radioactive waste	Air pollution	Water pollution (m3)	Acidification	Resource depletion	Water	Photochemical ozone formation
Total primary energy	**1.00**															
Renewable energy	0.77	**1.00**														
Non-renewal energy	**0.98**	0.61	**1.00**													
Primary energy process	**0.98**	0.69	**0.98**	**1.00**												
GHG	0.71	0.13	*0.84*	0.77	**1.00**											
Reused waste	0.07	0.02	0.08	0.05	0.25	**1.00**										
Hazardous waste	0.72	**0.95**	0.57	0.61	0.10	0.01	**1.00**									
Non-hazardous waste	0.14	0.02	0.16	0.08	0.16	−0.09	0.17	**1.00**								
Inert waste	0.39	−0.10	0.51	0.50	0.71	−0.03	−0.21	−0.10	**1.00**							
Radioactive waste	*0.84*	0.60	*0.84*	*0.83*	0.70	0.32	0.54	0.28	0.48	**1.00**						
Air pollution	0.72	0.77	0.63	0.72	0.36	0.16	0.64	−0.42	0.23	0.46	**1.00**					
Water pollution	0.58	0.71	0.63	0.56	0.16	−0.23	0.74	−0.11	−0.17	0.26	0.66	**1.00**				
Acidification	*0.83*	0.51	*0.86*	0.77	0.78	0.39	0.57	0.27	0.37	*0.83*	0.47	0.32	**1.00**			
Resource depletion	**0.93**	0.50	**0.98**	**0.95**	**0.90**	0.12	0.46	0.23	0.58	**0.88**	0.51	0.38	*0.86*	**1.00**		
Water	0.67	0.23	0.76	0.68	*0.85*	0.09	0.32	0.33	0.42	0.62	0.29	0.45	*0.80*	*0.80*	**1.00**	
Photochemical ozone formation	0.37	−0.15	0.51	0.35	0.71	0.17	−0.03	0.58	0.49	0.57	−0.26	−0.22	0.68	0.63	0.70	**1.00**

Bold very strong correlation: 0.90 to 1
Italic strong correlation: 0.80 to 0.90

We shall try to reduce the indicators around the one for GHG emissions.

The correlation of the 11 remaining products is:

Correlation between "GHG/resource depletion": 0.97

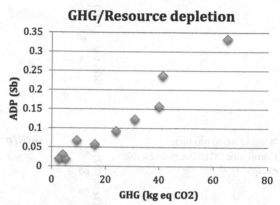

Correlation between "GES/water consumption": 0.85

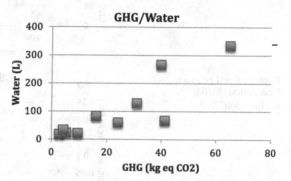

We can thus consider that the GHG indicator is correlated with the resource depletion indicator and with water consumption.

Once the "wood" products have been set aside, the correlation coefficients are the following:

Without wood product: Corrélation « **GES/énergie primaire** » est de 0.95

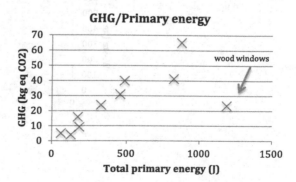

Without wood product:
Correlation "GHG/primary energy"
is 0.93

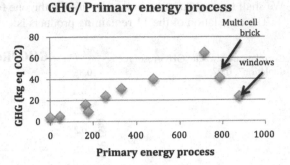

Without wood product:
Correlation "GHG/non-renewable
energy" 0.97

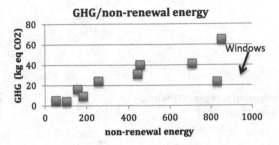

Without wood product:
Correlation "GHG/
acidification" is 0.88

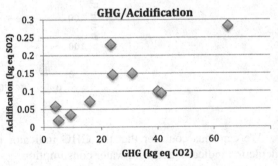

Excluding stone and brick multi cellular: Correlation of GHG and waste
recovered 0.86

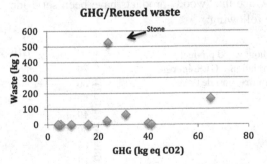

Without floor slabs and cellular concrete. Correlation emissions and radioactive waste: 0.90

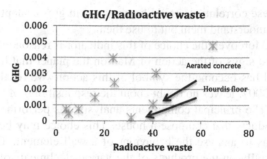

Without wood product and poured concrete correlation GHG/air pollution: 0.94

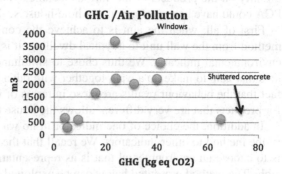

There is no apparent correlation between GHGs and hazardous waste, renewable energy, non-hazardous waste, inert waste or water pollution.

There is no correlation either between the "renewable energy" indicator and the others.

There is no correlation between the "non-hazardous waste" indicator and the others.

Finally, suppressing the plasterboard allows a correlation of 0.90 to be brought out between the hazardous waste indicator and acidification.

The following step is to carry out a reduction on the basis of a correlation in the values of the indicators.

Analysing the behaviour of the GHG indicator in our simulations allows the general behaviour of 9 of the 14 indicators making up the FDESs to be observed statistically with very acceptable sensitivity. After one to three "deviant" products out of the 12 selected have been set aside, the GHG indicator is correlated with 9 other indicators and the correlations are between 0.85 and 0.97. These indicators are:

- primary energy;
- process primary energy;
- non-renewable energy;
- recycled waste;
- radioactive waste;
- air pollution;
- acidification;
- resource depletion;
- water consumption.

These correlations could be analysed in greater depth but the objective here is not to understand them but to use them.

Moreover, the choice of this indicator is reasonable. It is one of the most widely used signs of the action of Man on the planet that has ever appeared in the media and has become the symbol of this action. We recall that the fourth GIEC report, in 2007, leaves no doubt about the responsibility of GHGs in climate change.

The principal components analysis was performed on the basis of the different products that compose a house. This choice may be criticized since the next step aims to analyse the behaviour of a wall element. The exercise could have relied initially on the products of the various technical solutions for the wall and a new PCA could have been performed at whole-house scale.

First of all, when the aim is to achieve the consistency needed to extend the method from the wall unit to a typical dwelling, it is indispensable to keep a single environmental indicator. We thus chose to simultaneously analyse different products that help to make up a wall, together with other components of the house. The fact that the behaviour results are close in terms of indicators, and thus of impact, for products that are very different allows the house to be analysed as a whole.

In addition, the choice of one indicator in no way compromises the application of the method to other indicators. We recall that the sole purpose of this first stage is to choose an environmental that is as representative of GHG emissions as possible. The method presented below can be exploited for:

- hazardous waste;
- renewable energy;
- non-hazardous waste;
- inert waste;
- water pollution.

Analysing the behaviour of all these indicators for the wall and then the whole building would provide a complete assessment of the impact of lifespan on the environment.

To sum up, studying GHG emission behaviour enables us to simultaneously study the GHGs, primary energy consumption, process primary energy, non-renewable energy, the production of recycled waste and radioactive waste, air pollution, acidification, resource depletion, and water consumption for the products.

5.3 Definition of the Element Under Study

- Analysis of needs and elements of functional specification (phase n°1)

The different solutions must correspond to the same functional specifications based on the usual standards (i.e. NF X50-151), summed up in the table below (Table 5.4):

Table 5.4 Specifications of the exterior wall

Réf	Functions	Requirements functions	Level criteria
FI1	Allow the recovery of all forces which it is subject	Masses Lateral forces wind Constraints due to openings	Conforms to EN 1990 Eurocode: 2002 Class A
FI2	Protect persons and property against the external elements	Resistance to wind Resistance to rain Resistance to cold Resistance to snow Allow summer comfort Acoustic attenuation (new acoustic regulations decree of 30 May 1996) Fire resistance Resistance to intrusion	Local level and EU: structural Eurocodes EN 1990 Depending on the local level local Attenuation > 50 dB (A) air Eurocode (see EN 1991-1-2) > A2P^{EUROCODE}
FI3	Allow the comfort and minimize impacts on the environment	Thermal resistance Limitations of air leakage Surface temperature and effusivity Summer comfort	$R = 3.7 \ \text{m}^2 \ \text{K} \ \text{W}^{-1}$ NF EN 13829 for the overall housing
FA1	Be pleasing to the eye and promote the desired sensation	Appearance/finishes Proprioceptors Memory	According to the local culture
FA2	Allow the creation of openings	Integration of openings	<2 m without recovery
FA3	Resist the external environment	Pollution Seismicity	Local Level
FA4	Ease of maintenance		
FA5	Participate in assets		

Constraints:

Economic
Technical
Regulations

The table below presents the functions

FI: Interaction function
FA: Adaptation function

The technical solutions proposed must fulfil these functions.

Fig. 5.1 Composition of the external wall

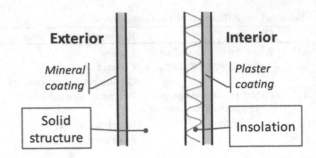

5.3.1 Chosen Solutions (Phase 2)

Various solutions corresponding to the same specifications have been chosen. All technical solutions for the external wall are composed of 4 layers, as shown on Fig. 5.1. The outside finish is a mineral coating and the inside is plasterboard. The insulation is glass wool, for which the only difference is the thickness required to provide the same thermal resistance, depending on the material of the solid structure.

The set of solutions correspond to the Fig. 5.1.

These technical solutions have the same thermal insulation coefficient, $R = 5$ m^2 K/W. It is impossible to act on the thermal insulation and the inertia simultaneously while guaranteeing conditions close to comfort. The inertia factor is all the more difficult to control because its effect on both winter and summer comfort is still poorly understood. The way inertia is considered in the evaluation tools needs to be improved (CETE Est 2010). When it is not possible to neutralize the differences in the physical characteristics of the technical solutions, the resulting variations in use should be integrated in the overall evaluation. Restrictive hypotheses are necessary given the limited way certain phenomena are taken into consideration, since finalized models are not yet available for them.

The technical solutions proposed should lead to reflection on the nature of the impacts these solutions may have on other elements of the system, and on the maintenance the choices impose if performance is to be kept up to the required level (Table 5.5).

We make the following assumptions:

- The "stone" solution is not covered with mortar coating as the main aim of mineral finishes is generally to look like stone. Although the great flexibility of products and the techniques for implementing them now affords variety in the aspect obtained, this restrictive hypothesis is maintained. There is no justification for coating a healthy stone wall.
- Strict compliance with the insulating function, neutralizing energy consumption during the operational phase, involves the addition of a thin layer of insulation in some cases. Even though this design is not realistic, it is kept in order to obtain the same thermal resistance for all solutions.
- Emission due to the vapour barrier is proportional to the thickness of the insulation.

Table 5.5 Technical solutions used for the wall

Material of the structure (thickness)	Insulation thickness (cm)
Stone (24 cm)	19
Concrete block (20 cm)	18
Wooden structure (23 cm)	19
Multi cell brick (30 cm)	10
Aerated concrete (20 cm)	6
Full solid clay brick (20 cm)	16
Mud	30
Shuttered concrete + PSE (20 + 8)	9

Table 5.6 Lifetimes for some of the material for a unit area (1 m^2)

Material	Mean lifetime (years) column 2	Lifetime interval (years) column 3	Comments
GW insulation	50	40–60	Industrials and scientists
Exterior rendering	30	25–35	Experts, Technical Director
Stone	750	500–1,000	On existing
Wood frame	70	60–80	Chief of cultural mission heritage Quebec/ Practitioner
Shuttered concrete	200	180–220	Concrete engineer
Full solid clay bricks	300	250–350	On existing
Concrete blocks	100	100	Builder
Mud	300	250–350	On existing
Aerated concrete	100	100	Builder
Multi cell bricks	100	100	Builder

Note although the principle of fuzzy logic theory accepts intervals for the lifetimes, for the sake of simplicity, the mean of each interval was taken for the calculations

- Convection and permeability phenomena are neglected.
- The walls have different thermal inertia but the choice of a temperate climate allows the energy consumed for summer comfort to be disregarded. This has been checked using the TRNSYS simulation program on a single-family house.
- The system for attaching the insulation is not considered because there are many different possibilities and the corresponding FDESs do not exist
- The impacts of thermal bridges, which vary from one solution to another, are neglected.
- The insulating material is always taken to be placed inside the building.

5.3.1.1 Evaluation of Element Lifetimes (Phase 3)

The hypotheses taken for the lifetimes are presented in the table below (Table 5.6).

Table 5.7   GHG emissions for all LCA

Material	GHG emission (Kg eq CO_2) column 1
GW insulation	0.309
Exterior rendering	5.13
Stone[1]	9.6
Wood frame	9.89 (*Ecoinvent*)[2]
	−18.27 (*INIES*)[3]
Shuttered concrete	65.10
Full solid clay bricks[4]	84.72
Concrete blocks	16
Aerated concrete	40
Mud[5]	7.5
Multi cell bricks	41.3

[1]Il est considéré que passé les 300 ans, la pierre continuera d'être utilisée. La valeur est donc calculée au prorata de la durée de vie
[2]Ce produit n'existant pas dans les bases INIES et KBOB, il a été l'objet d'un calcul par nos soins sur la base de produits existants
[3]Ce produit n'existant pas dans les bases INIES et KBOB, il a été l'objet d'un calcul par nos soins sur la base de produits existants
[4]Ce produit ne faisant pas parti de la base INIES, l'indicateur officiel n'existe pas. Nous empruntons la valeur à la base de données KBOB-ECOINVENT
[5]Indicateur non présenté dans INIES mais provenant du résultat provisoire du projet de recherche TERCRUSO qui porte sur les briques crues de la vallée de la Garonne

5.3.1.2 Definition of Data Used (Phase 4)

5.4 Environmental Data: Values of GHG Indicator

The values of the GHG indicators for the various products, evaluated in complete lifecycle analysis, are noted in Table 5.7.

Chapter 6
Results and Analyses

Abstract In this chapter, we show that it is indispensable to take the lifetime of a wall into consideration if responsible decisions are to be taken in the choice of technical solutions intended to optimize environmental performance.

Keywords Results • Sustainability development • Lifespan

6.1 Cumulative GHG Emissions for the Walls (Phase 5)

In this first phase, the calculations were performed from 0 to 300 years.

In terms of impact for a single material and according to the time factor, impact can be written as the function below:

$$\mathrm{Impk}(t_{j,k}) = (1+j) * a_k \quad \text{with } t_{j+1,k} = t_{j,k+dk} \quad \text{and } t_0 = 0$$

k   material number
Impk   Impact of the material solution (k)
a_k impact factor of the material (k) evaluated over the life cycle
d_k service life of the material (k)
$t_{j,k}$ time when the jth changes occur for solution (k)
j   time index.

For example, the service life of the mortar coating is set to 30 years, and the impact factor is 5.13 kg.eq.CO_2, so $d_k = 30$ and $a_k = 5.13$. At time 0 before the construction t_0, k = 0. Then, after the construction, at time $t_{1,k}$, $\mathrm{Impk}(t_{1,k}) = a_k$. Thirty years later $t_2 = t_1 + d_k$ and the coating is redone. Then, $\mathrm{Impk}(t_{2,k}) = 2 * a_k$, and so on.

$$\text{Then,} \quad \mathrm{Imps}(t_{j,s}) = \mathrm{Impk}(t_{j,k})$$

where s is the technical solution number.

Figure 6.1 shows the results for shuttered concrete wall (g), considering the hypothesis given above.

M. Méquignon and H. Ait Haddou, *Lifetime Environmental Impact of Buildings*,
SpringerBriefs in Applied Sciences and Technology, DOI: 10.1007/978-3-319-06641-7_6,
© The Author(s) 2014

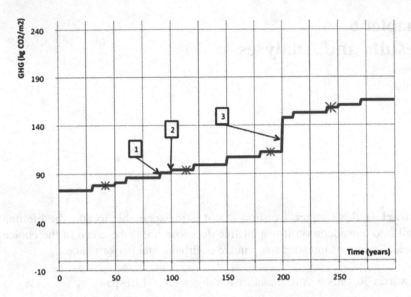

Fig. 6.1 Cumulative GHG emissions versus lifetime for shuttered concrete wall (g)

For the evolution of GHG emission during the expected lifetime of the wall, several "steps" can be observed on the curve of Fig. 6.1:

(1) represents the rise of GHG emission due to the renovation of the inside plaster coating, every 30 years.
(2) represents the rise due to the renovation of the glass wool insulation, every 50 years.
(3) is representative of the rebuilding of the wall since the service life of shuttered concrete is 200 years.

The impacts are calculated for whole life cycle analysis according to ISO 14040. To simplify, we considered that the impacts occurred during the construction phase. Using this principle, the curves for all technical solutions for walls are shown in Chart 6.1. The values of the results are given in Table 6.1 at 70 and 300 years.

The Chart 6.1 shows the evolution of GHG emissions for the different technical solutions. "Multi cell brick and Shuttered concrete", "Lifespan 70 years", "Wood solution (INIES)", "P Stone (Lifespan 300 years)" "Mud (Lifespan 300 years)".

Overall, the wall with the wood structure seems to be a successful solution, whatever the lifetime of the wall, but it depends strongly on the hypothesis taken for the GHG emission or absorption. These results show the contradiction between the two theories concerning wood. A solution to this question must be found quickly if consistent results are to be obtained.

Looking at the other technical solutions, the wall made with stone (a) is the best whatever the lifetime of the wall. This is because stone has the longest lifetime, between 500 and 1,000 years, and only the outer layers need to be changed over time.

Table 6.1   Results GHG emission for two lifetime; 70 and 300 years

Type of wall	nb	kg.eq.CO_2	
		70 years	300 years
Wooden structure (INIES index)	c	−5	1
Stone	a	32	57
Wooden structure (KBOB index)	c	25	107
Concrete block	b	48	148
Mud		35	111
Shuttered concrete	g	86	166
Full solid clay brick	f	114	178
Multi cell brick	d	65	200
Aerated concrete	e	65	211

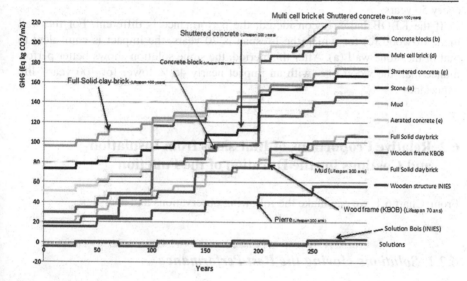

Chart 6.1   Evolution of GHG

Since the lifetime of a wall is never known with certainty in advance, this solution limits the consequences of an early demolition and deals with uncertainty on the basis of the precautionary principle. This solution seems to be the most sustainable and, compared to other solutions (wood INIES index not included), it allows a significant reduction, between 44 and 73 %, in emissions over a period of 300 years. This performance is achieved through the extremely long lifespan, the fact that energy requirements are restricted to cutting, transport and implementation, and also the absence of outside coating. In the case of rapid obsolescence, the stone can be easily reused. Considering the current state of technology and existing products, stone is probably the most recyclable solution.

These two solutions, stone and wood, which are the most effective because they do not use energy for transformation, are the least used, at least in France for new buildings. This is a paradox.

Focusing on the solutions having poorer performance, for a wall lifespan of less than 70 years, the "aerated concrete" (e) and "multi cell brick" (d) solutions give identical results and are better than shuttered concrete (g). Conversely, for a period longer than 200 years, shuttered concrete (g) is more efficient, with reductions of 17 and 21 % compared to aerated concrete and multi cell brick respectively . Thus, the relative performance of these solutions are closely related to their lifespans. It should also be noted that, for new buildings in France, the technical solutions (e) and (g) are the most widely used.

Note: The "wood solution" (c) as measured by the INIES index is the most effective solution whatever the duration of use of the wall. The negative value of the wood index due to long-term storage of CO_2 can offset greenhouse gas emissions by the mineral coating, renewed every 30 years, and the insulation, renewed every 50 years.

If the KBOB index is considered, the performance is different. For the wall made of wood (c) for which the lifespan is 70 years, the impact is equivalent to that of the stone wall (a). After this period, the stone solution shows better performance than the wood one, with an impact nearly 46 % lower for 300 years' life expectancy.

6.2 Relative Proportions of Emissions from Insulation and Coatings for the Duration of the Function

From Chart 6.1, we can make the following observations and analyses.

6.2.1 Solutions Having the Best Performance

The performance of the wood solution, as evaluated from the INIES index, is the best whatever the time for which the wall is in use. Its GHG emission completely disappears as the negative index of the wood (long-term CO_2 storage) compensates for the GHG emissions of the mineral coating (renewed every 30 years) and the insulating material (renewed every 50 years). This result should, however, be reconsidered in the light of the analysis given in Sect. 5.3, page XXX.

If we consider the KBOB index, the performance is different.[1]

Up to 70 years of use, this solution performs as well as stone. After this time, given the necessity for rebuilding, the performance of the wood solution is poorer than that of stone, with an impact that is 47 % higher at 300 years. Nevertheless,

---

[1] The results show the contradiction between the two principles used to calculate CO_2 storage for wood-based products. A consensus needs to be reached quickly if we intend to exploit the indicator values and obtain consistent results.

the "wood" solution maintains a relatively good position regarding performance whatever its lifetime, as it ranks second, practically at equality with the unbaked clay solution.

The "stone" solution gives good performance whatever the lifespan envisaged. Its emissions are strongly reduced, by between 47 and 72 %, relative to the others (except for wood in the INIES database), for a duration of 300 years. The architect Gilles Péraudin has presented stone as the ecological solution par excellence. For him, any stone construction can become a "quarry" for a later project. It reaches its performance thanks to its extremely long lifetime, its minimal energy requirements (restricted to cutting, transport and implementation), and the absence of added coating. Should the building become obsolete quickly, the stone can be reused easily. As techniques and products stand at present, stone is probably the most recyclable solution.

The "unbaked clay" solution, ranking joint second among the 7 solutions, behaves very well. Nevertheless, the maintenance and monitoring requirements imposed by its great sensitivity to humidity should not be forgotten.

Lower in the ranking, in the early years, we find "hollow concrete block". With the chosen the 100-year lifetime hypothesis, this solution always performs better then multi cell brick and solid baked clay brick. It reaches "average" performance for a service life of 300 years.

There is a paradox here. Stone, wood and unbaked clay, which rank among the solutions with the best performance in the comparison, are the least widely used, at least in France.

6.2.2 Second-Order Solutions

For a lifetime shorter than 70 years, the "aerated concrete" and "multi cellular brick" solutions, which have the same results, are preferable to the "shuttered concrete" and "solid brick" solutions. They reduce emissions by 25 and 43 % respectively. From 100 to 200 years, the aerated concrete, multi cellular brick and solid fired clay brick solutions emit the same amounts of GHGs. Beyond 200 years, shuttered concrete gives better performance, with reductions of 17 and 21 % relative to "multi cellular brick" and "aerated concrete" respectively. The "solid brick" solution provides reductions of 11 and 15 % compared to "multi cellular brick" and "aerated concrete" respectively. The relative performance levels of these solutions are thus closely linked with their lifetimes.

Table 6.2, which gives the absolute and relative emission value for each material in the walls, shows that, whatever the technical solution, the external coating has the same emission in absolute value, around 50 kg.eq.CO_2, as it needs to be replaced regularly. In relative value, for most of the solutions, it represents between 25 and 35 %, which is quite high. Note that the fact that the stone solution (a) does not need any coating explains why this solution has the lowest emission, if the wooden structure with the INIES index (c) is not taken into consideration.

Table 6.2 GHG emission for the different components of each wall, in absolute and relative value

Type of wall	nb	Total	Solid structure		Insulation		Mortar coating	
			kg.CO_2	%	kg.CO_2	%	kg.CO_2	%
Wooden structure (INIES index)	c	3	−78		37	–	44	–
Stone	a	59	7	12	52	88	–	–
Wooden structure (KBOB index)	c′	107	32	30	29	27	46	43
Mud		110	7	7	52	84	–	–
Shuttered concrete	g	183	98	53	34	19	51	28
Concrete block	b	151	49	32	52	34	50	34
Full solid clay brick	f	182	85	46	46	25	51	28
Multi cell brick	d	210	124	59	35	16	51	24
Aerated concrete	e	216	121	56	43	20	51	24

In most solutions, the solid structure is responsible for the majority of the emission, but the results are for 300 years and, if the wall lasts longer, then the proportion could be lower. For the wooden solution with the INIES index it is clear that the CO_2 absorption of the wood compensates, nearly exactly, the emissions of the other two layers.

The emission due to insulation varies from one solution to another but, as the thicknesses are different because they depend on the thermal properties of the solid structure, it is difficult to draw firm conclusions.

Finally, in most solutions except wood (c) and stone (a), the insulation and the external coating account for half the emissions. Even if the emissions during production are not very high, after 300 years, the global emission is significant because the service lives of these elements are too short: around 50 years for glass wool and 30 years for the coating. So, if the GHG emission from walls is to be lowered, efforts need to be made to develop more long-lasting solutions, especially for the layers mentioned.

6.3 Demonstration of the Uncertainties Inherent in the Official Indicators

Table 6.3 lists the solutions in order of decreasing technical performance as given by simulation, while Table 6.4 lists the technical solutions in order of decreasing performance as given by the official indicators.

A comparison of Tables 6.3 and 6.4 shows how decisions based on the official Functional Unit indicator, supposed to present the performance of a product, can be counterproductive. The mixed order of the figures in the green-shaded column provides a schematic indication of the errors in performance.

Table 6.3 Classification of technical solutions in ascending order of cumulative GHG emissions

Type of wall	Ranking	Eq.Co$_2$ output 300 ans	Official index
Wooden structure (INIES index)	1	3	−0.1076
Stone	2	62	0.1200
Mud	3	110	0.075
Wooden structure (KBOB index)	4	117	–
Concrete block	5	151	0.1600
Full Solid clay brick	6ème ex aequo	182	0.2640
Shuttered concrete	6ème ex aequo	183	0.651
Multi cell brick	7	210	0.2753
Aerated concrete	8	216	0.4040

Table 6.4 Classification of technical solutions in ascending order of official GHG indicators

Type of wall	Rank	Eq.Co$_2$ output 300 ans	Official index
Wooden structure INIES/KBOB	1 ou 4	3	−0.1076
Mud	3	110	0.0750[a]
Stone	2	62	0.1200
Concrete block	5	151	0.1600
Full Solid clay brick	6ème ex aequo	182	0.2640
Multi cell brick	7	210	0.2753
Aerated concrete	8	216	0.4040
Shuttered concrete	6ème ex aequo	149	0.650

[a]Rappel: Indicateur non présenté dans INIES mais provenant du résultat provisoire du projet de recherche TERCRUSO qui porte sur les briques crues de la vallée de la Garonne

6.4 Impact of Service Life of Technical Solutions on GHG Emission

6.4.1 Evolution of Emissions (Phase 6)

The Chart 6.2 shows the evolution of GHG emissions for each of the materials involved in the technical solutions according to the service of the solution.

The results can be analysed further. The interest of extending lifespans is obvious. As shown in Chart 6.2, setting the target at 100 or 150 years could lead to significantly improved performance and different choices. Whatever the solution, extending the lifespan from 50 to 100 years can reduce emissions by 50 % over a century. The extension to 300 years would allow a reduction of 83 % in emissions.

It is also established that a solution presented as having poorer performance may be just as good if the lifespan is proportionally longer.

For example, if we assume 100 kg.eq.CO$_2$, a wooden structure (KBOB index) built for 70 years has an impact similar to that of a wall made of stone if it is kept for about 50 years, hollow concrete blocks have the same impact over 100 years and solid clay bricks over 300 years. Whatever the level of emissions allowed, equivalencies can be found on the basis of different lifespans.

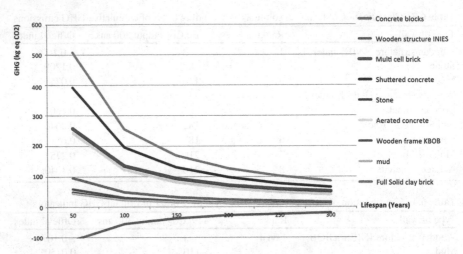

Chart 6.2 Evolutions GHG body according to lifespan solutions (The curves "Wooden frame KBOB" and "Mud" are superimposed)

The impact factor (a_k) is specific to the technical solution (k) and is directly related to the nature of the materials and the manufacturing process. Whether we consider the energy needed to produce the solution, or energy accounted for the complete cycle of the greenhouse gases emitted, or resource depletion, or any other environmental-impact or economic factor, this value is considered to be known, invariable data. In response to demand, manufacturers are seeking to reduce this factor technologically.

With constant performance, the results show that the impact of the wall unit is inversely proportional to its lifetime.

Note: The inverted curve of the "wood" solution based on the INIES index shows that this calculation cannot be used in the same manner. If it is used as for other products, it creates ambiguity or even real misunderstandings. Indeed, it would lead to an absurd paradox: "The longer the lifespan is, the less favourable is the impact in terms of GHG". This would be an encouragement to reduce lifespan to a few years, without any thought for the waste due to the demolition … It seems that this negative value is caused by confusion in the wood characteristics. It is necessary to examine these specific characteristics and provide consistent information that will show the performance of all wood products according to their use.

To reduce the emission of greenhouse gases, it is better to extend the lifetime of a solution rather than choose the lowest impact during the construction phase. To find optimal solutions, of course, the strategies in the choice of technical solution and the increase of the lifespan can be accumulated. EUROCODE 0 sets life expectancies for calculating the sizing of buildings based on statistical estimations of failure. Although a lifespan of 50 years is used for ordinary buildings and has become an objective for professionals, it is clear that it does not encourage optimization in terms of GHG emissions.

Generally speaking, the amounts of natural resources consumed and GHGs emitted for an element to fulfil the function it is intended for decrease in inverse

Table 6.5 Lifespan technical solutions for equivalent emissions

Technical solution	Trend equation (column no 2)[a]	Duration equivalent (years) (column no 3)
Full solid clay brick	$y = 508.3x^{-1}$	339
Shuttered concrete	$y = 390.6x^{-1}$	260
Multi cell brick	$y = 253.6x^{-1}$	194
Aerated concrete	$y = 242.4x^{-1}$	162
Concrete block	$y = 96x^{-1}$	64
Wooden structure (KBOB index)	$y = 59.32x^{-1}$	40
Stone	$y = 57.6x^{-1}$	39
Mud	$y = 45x^{-1}$	30

[a]The calculation requires the change of variable formula. If X is the number of years, X = 50 x

proportion to the increase in lifetime of the element. For example, and very logically, if the utilization function is required for 300 years, the consumption of aggregate granulate will be six times smaller if the product service life was 300 years instead of 50 years. Conversely, if the element is destroyed earlier than expected, the consumption per unit time will show a proportional increase.

Note: This observation is true for all materials, and so all the impacts, whatever the solution chosen.

6.5   Comparison of Technical Solutions

The trends in GHG emission behaviour for the various solutions according to their lifespans are given by the equations noted in column 2 of the Table 6.5. For an equivalent emission of GHG, e.g. 75 kg.eq.CO_2/m^2, considering a utilization function of 300 years, the time each solution needs to last can be found in column 3 of the table.

It is established that a solution presented as having lower performance can be just as good as another on condition that its lifetime is proportionally longer.

For example, if a value of 75 kg.eq.CO_2/m^2/300 years is fixed for emissions, equivalence of impact will be obtained if the actual lifetimes take the following values.

A stone wall must last 39 years. An unbaked earth wall must last 30 years. A wooden wall (KBOB index) must last 40 years, a wall of hollow concrete blocks 64 years, one in aerated concrete 162 years, in multi cell bricks 194 years, in shuttered concrete 260 years, and in solid clay bricks 339 years. Whatever the level of emissions assumed, equivalencies can be found on the basis of different lifetimes cf. Table 6.5.

Note: The inverted curve for the "wood" solution, base on the INIES index, clearly shows that this index cannot be used in the same way as the indices for other products. As it stands, and used as we have done for the other products, it creates ambiguity and can even result in errors. In fact, it leads to the absurdity that "the longer the lifetime, the less favourable the impact in terms of GHGs".

So it would be worthwhile to reduce the lifetime to a few years or, even better, a few days without any conditions being fixed for the fate of the demolition waste ... It seems that this negative value comes from confusion of ideas (storage during utilization and fate after utilization) for the characteristics of the wood. It thus appears necessary to look more closely at these specific characteristics and supply consistent information presenting the interesting performance levels of wood products.

6.6 Conclusion

It is indispensable to take the lifetime (explicit or not) of a wall into consideration if responsible decisions are to be taken in the choice of technical solutions intended to optimize environmental performance.

Concerning GHG emissions and the 9 other indices with closely related behaviour, the effective lifetime of construction products has an effect on the environmental consequences of the way a utilization function is satisfied for a given duration. For example, the "hollow concrete block" solution would be penalized by a service life of 100 years, whereas its performance would be among the best for longer lifetimes. The actual lifetime is a very important performance criterion.

The official indicators, based on the functional unit value, can prove counterproductive in a design approach aimed at reducing impacts. It is not the evaluation of the indicators made in life cycle analysis that is in question here but the typical lifetime.

To compare two products by impact factor, comparing the impact factors in isolation or in association with a fixed time factor would be a mistake. It is absolutely necessary to chose the time factor closest to reality.

The factor that needs to be understood in greater depth is the service life.

The products used for coating and insulation are not negligible in terms of GHG impacts. In parallel with efforts to reduce emissions during the utilization phase, it is important to increase the lifetimes of these two elements. Their service lives should thus also be studied with a view to reducing impacts.

The indicators of the various solutions could be compared but the objective of this work is not to present the solution for the wall unit that performs best with respect to the various criteria but simply to analyse the impact of the lifetime factor.

Finally, in contrast, there are a few positive aspects that result from reduced lifetimes:

• Improvements that come with experience. The shorter the lifetimes of products and complete life cycles, the greater the number of cycles—and each cycle enables us to improve our knowledge through experience.
• Possible improvements in performance connected with technological progress. The technological evolution of manufacturing processes and of the products themselves brings hope of gains through the use of new products.

Part III
Conditions for Generalization and Prospects

The results bring to light a need for deeper understanding and the extension of various aspects.

Chapter 7
Conditions for Generalizing the Approach

Abstract This chapter sets out the conditions of the generalization of the method and results to other building components.

Keywords Limits • Modelisation

7.1 Analysis

The approach is not at all difficult to generalize. Only the issue of the links existing within the system can lead to a chain of end-of-life events. Floor and partition wall coverings will have no consequences for the lifetimes of other elements but, in contrast, the end of lifetime of the main walls will entail the dismantling or even the demolition of the rafters, roofing, electric circuitry, plumbing, etc.

The end of life of the foundations is the event most likely to lead to the demolition of the other elements as the probability of rebuilding foundations in the same place is low.

7.2 Modelling Elements

7.2.1 Evolution of the Impact of a Product i Over Time

From these results we can deduce how the impact behaves as a function of time. We can write the function as:

$$\mathrm{Im}\, p_i(t_j) : \begin{array}{c} [t_j, t_{j+1}] \\ t_j \end{array} \quad \begin{array}{c} \mathrm{IR} \\ \mapsto (1+j)a_i \end{array} \tag{7.1}$$

M. Méquignon and H. Ait Haddou, *Lifetime Environmental Impact of Buildings*,
SpringerBriefs in Applied Sciences and Technology, DOI: 10.1007/978-3-319-06641-7_7,

such that, $t_{j+1} = t_j + d_i$ with $t_0 = 0$

Imp_i Impact of technical solution i according to the service life of the technical solution

a_i Impact factor of solution i evaluated over the complete life cycle of the technical solution or product

d_i Service life of technical solution i

t_j Time step

Example Impact function of mineral rendering (noted me) for which the service life is $d_i = 30$ years, relative to its time in use.

$$\text{For } j = 0 : t_{0+1} = t_1 = t_0 + 30 = 30 \quad Imp_{me}(t_0) = 1 * a_i = 5.13$$
$$\text{For } j = 1 : t_{1+1} = t_1 + 30 = 60 \qquad Imp_{me}(t_1) = 2 * a_i = 10.26$$
$$\text{For } j = 2 : t_{2+1} = t_2 + 30 = 90 \qquad Imp_{me}(t_2) = 3 * a_i = 15.39$$
$$\vdots \qquad\qquad \vdots \qquad\qquad \vdots$$

7.2.2 Evolution of the Impact of a Product i According to Its Service Life

The impact of a product according to its service life is represented by the function

$$Imp_i(d_i) = \frac{a_i}{d_i} \times D \tag{7.2}$$

with

Imp_i Impact of the technical solution i according to its lifetime

a_i Impact factor of solution i evaluated over the complete life cycle of the technical solution

d_i Service life of technical solution i

D Duration of the service function

   Equation (7.2) above shows that the impact in terms of sustainable development is inversely proportional to Di. A short lifetime generates a large impact.

7.2.3 Analyses

The impact factor "a" is the factor attached to technical solution i itself. This factor is directly connected with the nature of the materials and the manufacturing process. Whether it concerns the energy needed to produce the technical solution, the energy used over the complete life cycle, the greenhouse gases emitted, resource depletion, or any other environmental or economic impact factor, this value constitutes data. In response to widespread demand, industrial firms are trying to reduce the factor through technology.

For a given duration, we obtain one impact value. For a function such as housing, considering the need at human scale, this duration must, by definition, be long.

The variable of the function, the service life, d_i, of product i is, by definition, the factor that makes the function inversely proportional to time. This factor has not yet been specifically and precisely studied by scientists.

7.2.4 Impact of Service Life and Technological Improvements

In the context of choosing between demolition/rebuilding and renovation, it is necessary to consider, on the one hand, technological progress and the impacts associated with the new solution and, on the other, the impacts connected with the renovation and maintenance of the initial solution. We recall that, in our work, the performance of the solutions was maintained by replacing the insulation every 50 years and the protective outside coating every 30 years. These replacements were, of course, taken into account in the calculations. Thus the impacts of the maintenance needed for the upkeep of the wall unit are included and integrated in the calculations. They are thus taken into consideration in the strategic decision process to choose between keeping up the function or dismantling and rebuilding. The fact that the lifetime of the "load-bearing function" of our wall unit is lengthened does not mean that we should not look for ways to improve the other functions, such as insulation.

7.2.5 Comparison of Products

From Eq. 7.2, established earlier, comparing two products, 1 and 2, that provide the same service in terms of greenhouse gas impact for a length of time D imposes a comparison between

$$\frac{a_1}{d_1} \times D \text{ et } \frac{a_2}{d_2} \times D$$

a_1 will really give better performance than a_2 if and only if:

$$\frac{a_1}{d_1} < \frac{a_2}{d_2}$$

So it is absolutely necessary to know d_1 and d_2.

We recall that the comparisons are made for identical service rendered and so, taking a few reducing hypotheses, for identical impact in use. Let us not forget that the final choice between two products, based on integrated performance in terms of sustainable development, would require the use of multi-criterion tools.

Chapter 8
Limits

Abstract This chapter outlines the limits of the results. The limits of these results come from the fact that they do not take the technological innovation factor into consideration. True, the construction field, with the exception of specific constructions such as buildings in earthquake zones, is not very sensitive to technological evolution.

Keywords Lifespan • Limits • Technical solution

The biophysical aspect of the principal elements of comfort is neutralized among the various solutions. Having the same insulation coefficient and inside and outside materials neutralizes losses, surface temperature and effusivity. However, the moisture transfer, internal heat convection and phase difference aspects are neglected. The psychological aspects of comfort and mood associated with the use of solutions that are more or less harmful for the environment are also neglected.

Studying a wall unit does not enable the impact of lifetime on the system to be envisaged for the various technical solutions, e.g. heat losses per unit length. Neither does it allow the impact of the summer comfort of the various solutions to be included as the thermal insulation and inertia parameters are difficult to neutralize at the same time. To study these two parameters simultaneously, it is necessary to make local simulations in order to compare technical solution performance levels.

The differences in the acoustic performance of the solutions are neglected.

The reductions of impact made possible by technological improvements in the product manufacturing or implementation processes have been neglected. The impacts when the products are renewed have been taken to be constant.

The 300-year duration of the housing function, on which the simulations are based, is justified by observations on existing residential buildings. It is obvious that, whatever the technical solution, the duration of its service life is limited. Three hundred years may be a utopian duration for certain solutions. There is probably a lifetime beyond which the economic and environmental costs of maintaining a service would cause replacement to be preferred in a context of optimization. To consider

M. Méquignon and H. Ait Haddou, *Lifetime Environmental Impact of Buildings,*   99
SpringerBriefs in Applied Sciences and Technology, DOI: 10.1007/978-3-319-06641-7_8,
© The Author(s) 2014

this question effectively, it is necessary to know both the physical lifetimes of the solutions and the conditions required for them to be attained.

Only the walls of residential buildings have been analysed in this work. It is probable that buildings intended for other uses, which may be subjected to greater mechanical stress making increasingly expensive upkeep necessary within a short timeframe, would obey different laws of behaviour.

Other impacts on the environment, such as hazardous waste, renewable energy, non-hazardous waste, inert waste and water pollution, should be evaluated through the question of the influence of service life duration. The conclusions for the performance levels of the technical solutions are valid for the indicators chosen. The performance of solutions can sometimes be contradictory. This may be the case, for example, for a renovation that would be more expensive than demolition/reconstruction and yet would be preferable from the environmental point of view. Solutions evaluated by means of different indicators can thus show contradictions in performance that prevent any conclusion being drawn on the basis of a comparison of technical solutions. A comparison of technical solutions imposes the use of multi-criterion decision-support tools in this case.

Without calling the trends into question, these results are limited in time as frequent changes in indicator values make it necessary for the information used in evaluations and comparisons to be updated in real time. This limit also applies to tools for evaluating product performance. How can we hope to encourage manufacturers to improve the performance of their products if the data used by evaluation tools do not keep pace?

The limits of these results also stem from the fact that they do not take the technological innovation factor into consideration. True, the construction field, with the exception of specific constructions such as buildings in earthquake zones, is not very sensitive to technological evolution. Its sensitivity is more noticeable for questions concerning equipment, which can be accommodated without making any great demands on the structure. It should also not be forgotten that the results are those obtained using knowledge that is currently available.

In this report we have not considered the risk of lifetimes being reduced by accidents or the possibility of decisions being taken that would cause the utilization of a building to be prolonged.

Chapter 9
Interest and Prospects

Abstract This chapter presents the contributions and the interest of our approach for improving the actual performance in terms of sustainable development.

Keywords Sustainable development approach • GHG • Primary energy • Optimization

9.1 Contributions to a Sustainable Development Approach

The importance of lifetime has been demonstrated as far as environmental impact is concerned. At least for certain indicators, the impact of service life on sustainable development is real. The study of the GHG indicator enabled the actual behaviour to be simultaneously revealed for primary energy, process primary energy, non-renewable energy, recycled waste, radioactive waste, air pollution, acidification, resource depletion, and water consumption. More precisely, whatever the technical solution chosen, a short lifetime for a residential building does not generally optimize performance in an environmental approach. Lengthening the lifetime always seems more favourable. Conversely, when the lifetime is fixed, a priori, the information can guide the choice of technical solutions with a view to optimization. Once the service lives of the technical solutions are better known, the designer will be able to choose the solution having the least impact on the selected criterion according to the expected lifetime of the building.

The economic and social impacts of lifetime need further study. By definition, a sustainable development approach consists of avoiding actions that can be harmful to future generations. Ancient buildings are very often positive signs bequeathed to us by previous generations.

We have shown that, for the greenhouse gas indicator, the lifetime of a product in situation is more important than the choice of product itself. The results imply that we should not observe only the raw performance of a product to decide whether it is environmentally friendly or not. It is necessary to know its intrinsic, theoretical service life and also the behaviour it will induce in its users and owners, which will have repercussions on its actual lifetime.

Energy savings made, or at least theoretically possible, during the utilization phase have caused the relative importance of the production/demolition phase to increase. And yet studies of the impact of production/demolition are rare in comparison with works concerning the utilization phase.

EUROCODE 0 defines durations for building sizing calculations based on failure statistics. Although it only concerns statistics in the field of mechanics and is not intended to specify building lifetimes, it has been shown that, if the 50-year lifetime mentioned for ordinary buildings became an objective in the profession, building performance in terms of sustainable development would deteriorate very seriously.

In addition, the FDES, which are very interesting documents in many respects, mention typical lifetimes that are misleading. For the information to be pertinent, it would have to be assumed that:

- the product does not influence the behaviour of the user or owner
- all the products have the same physical lifespan.

Calling these typical lifetimes into question leads us to question the information relating to functional units. The only indicator values that can really be exploited are those evaluated in complete life cycle analysis, phase by phase or globally.

The tools for evaluating performance in terms of sustainable development do not take lifetimes into consideration in a pertinent way. On the basis of fixed, theoretical or "typical" lifetimes, these tools give results that may be far removed from the actual situation. Moreover, if no warning is provided concerning the importance of the lifetime parameter, they may well lead to decisions having counterproductive effects. It is thus necessary to organize serious thinking about the methodology for taking lifetime into account.

Finally, our study focuses on the impact of lifetimes on greenhouse gases. The behaviour brought to light needs to be valid for all emissions.

9.2 Research Prospects

9.2.1 Impact of Lifetime for a Building

The demonstration made here for a wall can be envisaged for a building as a whole. Studying the impact of the lifetime on the environment will shed light on certain realities. Studying the impact on the building will, among other things:

- bring out the relative effects of the construction/deconstruction and utilization phases;
- verify that the impact of winter comfort has been neutralized and allow the differences of impact of summer comfort to be measured;
- render the differences of impact of extreme solutions commensurable and easier to understand;
- enable gains due to the extension of a building's lifetime to be measured;
- permit the social impact of the length of the building's lifetime to be analysed.

9.2.2 Multidisciplinary Reflection for a Sustainable Development Approach

It also appears necessary to organize multidisciplinary reflection on the characteristics of a building respecting a sustainable development approach. While a relative consensus seems to have been reached on the environmental approach, the economic and social approaches have yet to attain such a degree of maturity. Many actions and much thought are currently oriented in this direction. We note the development of many actions aimed at establishing a multidisciplinary approach with a view to improving the efficiency of the choices made in the building domain, some of which will doubtless result in the publication of works that will clarify the contents of an integrated approach in terms of sustainable development.

9.2.3 Indicators and Evaluation Tools

There is a need to develop indicators and integrated tools for evaluating performance in terms of sustainable development to help in decision making. Studies should be undertaken to define the possibilities of integrating environmental, economic and social indicators. The definition of such integrated indicators will require reflection based on analysis that will enable tools to be developed to support the decision-making process. These tools need to implement multi-criterion choices. Finally, the tools for evaluating performance in terms of sustainable development should consider the lifetime question more efficiently.

9.2.4 Deeper Understanding of the Relative Contributions of the Component Parts of a Building

The study of the relative contributions of the various component parts of a building to the impacts of the approach in terms of sustainable development would make for greater efficiency of action and more efficient use of the means employed in the investigations.

9.2.5 Recycling

The amount of recycling and the possibilities of reusing materials still appear to be small in the building sector. And yet the efficiency of this characteristic is a decisive factor in performance and impacts as far as sustainable development is concerned. Improving understanding in this domain implies obtaining greater in-depth knowledge of the difficulties and gains that can be expected of the actions. How much would service life need to be extended to compensate for a smaller recycling capacity? What materials and building systems are favourable in this domain? What is the performance status of the various recycling channels? What networks exist for professional waste collection? What know-how do the professionals possess? What are the costs and the gains?

This work has attempted to study the consequences of a building's lifetime for sustainable development. However, it is also important to study the causes that may affect this lifetime:

- the causes influencing the lifetime of buildings;
- the characteristics of a building that are likely to enable it to adapt to evolving needs and thus delay its obsolescence;
- the correlation between the lifetimes of buildings and the quality of their integration in their environment;
- the importance the project manager attaches to various priorities during construction or his decisions as to the possibilities of renovation or demolition;

9.2.6 Adaptability

Building designers and project managers should have the possibility of knowing the aspects that favour adaptability. This knowledge is necessary to enable them to take the measure of the overall cost and the interest of different solutions. Such a requirement implies the need to study the characteristics that have allowed certain buildings to become long lived by adapting continuously or at specific points in time according to needs. The knowledge acquired needs to be projected over a long time and the difficulties linked to uncertainty on the data and on future constraints mean that designers and decision makers must have high levels of expertise and awareness.

9.2.7 Total Number of Buildings and Demand

It is necessary to study the existence of links between population density and the lifetime of buildings, and also the impact on buildings for housing. The lifetime of buildings obviously has an effect on the total number of buildings available (cf works by N. Kohler; Van Nunen 2010, 2011, etc.). Whether it is extended intentionally or not, a longer building lifetime increases the stock of available housing, and inversely. In addition, quantitative needs, at least those corresponding to the primary function in Maslow's sense, are proportional to the population status. Population forecasts exist, and could allow simulations to be made in order to assess demand, to study probable needs and to calculate the number of dwellings necessary according to their lifespan.

9.2.8 Impact at Different Scales

The question of the impact of lifetime on sustainable development should also be widened though a study of:

- the connections among a building, its neighbourhood, the town, the urban area and the general area. A building is not just a product like any other. It is probably the object that goes furthest beyond the link with its owner and its user by dominantly imposing itself upon other individuals, passers by and the whole of society. Moreover, this specific relationship is a lasting one. It is the element whose associations generate districts and, through them, cities. Have the relationships thus fostered been examined through the prism of sustainable development?
- buildings in other sectors. Do service, commercial and industrial buildings show similar behaviour in the impact of their service lives?

To precisely evaluate performance in terms of sustainable development, the technician necessarily has to acquire better knowledge of the service life of the products used in situation. It is thus important to:

- carry out technically and scientifically established evaluations of product life times in normal conditions of use;
- study the consequences of any prolongation of use not noted in the initial programme. For this, it is necessary to set up evaluation tools on the basis of the factors listed previously;
- study the consequences of accidents occurring that lead to deviation from the initial expectations;
- gain deeper understanding of the issue of contradictory information concerning products based on plant materials. This information, founded on points of view that differ drastically from one database to another, engenders uncertainties and can lead to counterproductive decisions. These materials are of great interest in as much as they are used for their quality. It thus appears urgent to look for a way to take advantage of their qualities at different stages of their life cycle. Can we really imagine that the CO_2 they have stored is not released at the end of their lifetime? In contrast, assessment over the complete life cycle is not able to bring out the storage function. How can its advantages be highlighted?

References

Van Nunen H (2010) Assessment of the sustainability of flexible building. The improved factor method: service life prediction of buildings in the Netherlands, applied to life cycle assessment. Eindhoven 2010

Van Nunen H (2011) Improved service life predictions for better life cycle assessments. In: Proceedings of SB11 world sustainable building conference

Conclusion

The service life issue is frequently mentioned, including in construction standards. Nevertheless its impact on performance, be it environmental, economic or social, had never been evaluated with precision. We hope that the present work will have convinced researchers and decision makers that taking account of the instantaneous performance alone has every chance of engendering a result opposite to the one desired as far as the environment is concerned. It remains to be verified that performances do not oppose one another depending on the criteria applied, as such opposition would impose the need to rank the impacts in a hierarchy and also to develop multi-criterion tools.

In our conversations with various researchers, the question of building lifetimes was sometimes perceived as trivial through answers imagined immediately and logically, sometimes rejected because of the complexity it presented when it was envisaged mechanically and statistically, or sometimes evoked for the seriousness of its possible results in terms of responsibilities. At no time has our work rejected these points of view, which are all perfectly admissible. The objective of this work has been to spark reflection and produce knowledge on the basis of technical arguments capable of changing behaviour patterns in the approach to construction projects. We hope we have contributed to the idea that, on the one hand, service life has direct impact on performance in terms of sustainable development and, on the other, this factor is too often not known or is neglected in the work performed and the tools developed.

In the field of product design and production, buildings have a special status in the sense that they transcend the technical objects in which they are embodied. A building is one of the vectors, or possibly the vector, of communication with future generations, carrying messages about the most sustainable lifestyle, culture and principles of society. This reflection, already in train, should also be extended to questions on the existence, the nature and the objectives of the messages we want to transmit to the generations to come.

If we consider the definition of sustainable development to include the linking of responsibilities between generations, we are all too often unaware of the

M. Méquignon and H. Ait Haddou, *Lifetime Environmental Impact of Buildings*,
SpringerBriefs in Applied Sciences and Technology, DOI: 10.1007/978-3-319-06641-7,
© The Author(s) 2014

questions of time and the lifespan of our built objects. Our question could be posed for all other sectors of industrial or craft activity. What is the impact of the lifespan of consumer products on sustainable development? It seems to us that the lifetime of the objects we are surrounded by is a fundamental issue in the responsibility each generation has to itself and to others. The service life of goods, which are produced by consuming resources, has a direct effect on the volume of resources required to satisfy a functional need. The shorter this lifetime, the greater the amount of resources consumed and the smaller the amount available for other functions, which may be more important. Admittedly, reductions in resource consumption made possible by technological advances must enter into the equation ... and we are faced with a question of society in which the technician has to play a role.

Appendix 1
Référentiels

M. Méquignon and H. Ait Haddou, *Lifetime Environmental Impact of Buildings*,
SpringerBriefs in Applied Sciences and Technology, DOI: 10.1007/978-3-319-06641-7,
© The Author(s) 2014

Appendix 1 Référentiels

Title	Description	Qualities	Faults	Take account the lifespan
HQE®/QEB France	14 targets: New Buildings/Rehabilitation	- Flexible System - Encouraging adaptability	Lack of precision	No consideration of the impact on the life indicators DD
http://assohqe.org/hqe/	Relationship building with its environment			No enhancement of adaptability
	Integrated choice of products, systems and construction processes			
	Chantier with low environmental impact			
	Energy management			
	Water management			
	Waste management activity			
	Maintenance and sustainability of environ- mental performance			
	Hygrothermal comfort			
	Acoustic comfort			
	Visual comfort			
	Olfactory comfort			
	Sanitary quality of areas			
	Health Air Quality			
	Sanitary quality of water			

(continued)

Appendix 1 (continued)

Title	Description	Qualities	Faults	Take account the lifespan
LEED/USGBC Canada http://www.usgbc.org/	- Includes a list of items to take into account the environment and "win" unit credits; - Aimed at professionals and communities; - New buildings, Commercial and service buildings, neighborhoods réhabilitation; - Optimize Energy Performance valuation: - Protection of the Ozone Layer (HCFC-free materials) - Reduction of waste, recyclable materials - Reuse of existing materials-based recycling, easily recyclable, - Cost 300 to $ 400	- Reference manual that outlines all Canadian requirements - Enables certification projects and professional accreditations - Taking into account the effect of heat island with respect to proximity.	- Weighting of criteria	No real impact assessment of the lifespan, no weighting of the criteria life extender relative to other criteria
BREEAM UK www.breeam.org	New buildings Building renovation Neighbourhood Development Management Health/Wellness Energy Transport Water Materials Land use and ecology Pollution Innovation	Global cost assessment	- No social and cultural assessment - No used stage assessment	No consideration of the impact of lifespan on sustainable development indicators

(continued)

Appendix 1 (continued)

Title	Description	Qualities	Faults	Take account the lifespan
CASBEE Japon http://www.ibec. or.jp/CASBEE/ english/	Tool with multiple modules design assistance, for new construction, existing construction management, renovation and urban the delivery of certification	- Energy, resources and materials assessment, - External and inside environment, quality of Service;	- No comparison design solutions - No study and economic indicator	- Extending the life of maintenance through is taken into account. However, the precision required to limit to 30, 60 or 90 years and the life of the housing; - Lifespan are fixed; - No specific study to optimize lifespan; - The impact of life does not seem to have been a thorough case study.
	CASBEE evaluates the Q/L ratio with Q for Building Environmental Quality and Performance & L Building Environmental Loadings - Weighting system: The weights of evaluation fields are not determined solely by scientific data, they also take into account the values and perceptions of different stakeholders such as designers, owners, managers. The rating scale is from 1 and 5	- Calculates GHG full life cycle - Taking into account the building flexibility		

(continued)

Appendix 1 (continued)

Title	Description	Qualities	Faults	Take account the lifespan
HK-BEAM Hong-Kong	- Assessment tool of environmental impacts for residential buildings, commercial building and institutional buildings. - Associated with a standard and certification	- Takes into account the re-use and adaptability - Credit winned for reducing the impact of air conditioning - Taking into account the costs and the social and cultural impact - Tables calculations provided Takes into account: - The location of the site planning and programming - Use and Waste Management - Energy consumption - Water consumption and thermal comfort, indoor air quality lighting, noise, vibration … - Modular development leading to the evaluation unit in the form of credit		Lifespan existed without development

(continued)

Appendix 1 (continued)

Title	Description	Qualities	Faults	Take account the lifespan
Green Globes Tools www.greenglobes.com US/Canada	Assessment of the environmental impact of new or major renovation in the form of modules providing points and a score for the overall result constructions.	- Easy to use	- No visibility in the choice of the weighting of modules - Any consideration of the lifespan - The latest version establishes a global cost approach.	For the consideration of lifespan: - Adaptability of construction: 5/1000 - Reuse of existing 15/1000 - Impact of the life cycle of materials and the system as a whole 40/1000 - No serious consideration of the lifespan on impact on indices DD
Céquami http://www.mamaisoncertifiee.com/	Organized by building professionals that support the improvement of a house after initial assessment of the existing house in accordance with the requirements of the Standards	Certification guarantees: - Occupant Safety - Energy Efficiency - Comfort - Reduction of health risks - Water Management - Materials and equipment - Reduced impact of construction - Conservation of certification		No consideration for lifespan

(continued)

Appendix 1 (continued)

Title	Description	Qualities	Faults	Take account the lifespan
Passiv'Haus® Allemagne	is a German label energy performance in buildings. It is given to new homes with heating needs are less than 15 kWh/m²/year.	Concerne les questions énergétiques	Pas d'évaluation de développement durable	No consideration for lifespan
MINERGIE® ou MINERGIE-P® MINERGI-ECO(Suisse) www.minergie.fr	is a complement to MINERGIE ® or MINERGIE-P ® - Applies to new and rehabilitation	Respect to energy issues	No sustainable development assessment	No consideration for lifespan
Eco-Pro Habitat Belgium	- Issue of certificate following the respect of a charter	Simple and Free		No consideration for lifespan

Appendix 2
Tools of Sustainable Development Assessment

M. Méquignon and H. Ait Haddou, *Lifetime Environmental Impact of Buildings*,
SpringerBriefs in Applied Sciences and Technology, DOI: 10.1007/978-3-319-06641-7,
© The Author(s) 2014

Appendix 2 Tools of Sustainable Development Assessment

Title	Description	Qualities	Faults	Take account the lifespan
ELODIE-CSTB	Calculations of environmental impacts in terms of: - Depletion of resources - Pollutants - Noise in the neighborhood	Calculates the environmental impacts of building products from INIES	- Assessment tools specifically on environmental issues - Taking into account the typical lifespan products	- Taking into account the typical lifespan products
PAPOOSE-France	Tool for decision support targeting owners, developed by BE TRIBE concerns the design phase	Calculations of 13 environmental indicators	Considers the economic aspects	- Duration of lifespan fixed
Eco-Quantum Pays-Bas	Calculation of the environmental effectiveness of construction plans and new energy systems, efficiency and cost improvements to the environment. Focuses exclusively on residential buildings.			Lifespan: 50 years
Envest 2 UK http://envestv2.bre. co.uk/	Evaluation method as eco-points with an overall score to measure the environmental and economic performance of the building and the block		–	

(continued)

Appendix 2 (continued)

Title	Description	Qualities	Faults	Take account the lifespan
IISBE/SBTOOL/-GBTOOL Canada www.sbis.info	Modular development leading to the evaluation from 1 to 5 points with a final score Lead to certification project Module specifications local priorities Designed for new construction and renovations large commercial buildings and residential	- Adaptation of the tool to the specific local - Taking into account the costs and the social and cultural impact - Taking into account the flexibility - Taking into account local priorities. - Evolutionary tool based on project phases - Precise tool, only to adapt to local constraints and priorities. - Only to consider cost issues, social and cultural ways View summary E4, F1, F2, G1 and G2 - Rating: Deficient (-1) Good (0) Good (3) Best (5)	- The parameters of the economic assessment are left to the discretion of the user who must have expertise in the field - Evaluation at the end of construction.	- No real impact assessment of the lifespan
ATHENA Eco-calculator USA/CANADA	Assessment of the impact of building materials (manufacturing, transportation, construction, maintenance, demolition)	Very synthetic tool	Failure to take into account the costs No analysis of the impact of life on SD indicators	No analysis of the impact of lifespan on indicators of sustainable development

(continued)

Appendix 2 (continued)

Title	Description	Qualities	Faults	Take account the lifespan
EQUER	EQUER is an assessment tool of the environmental quality of buildings, designed to help stakeholders to better understand the consequences of their choices. As an analytical tool is used by all building professionals. An architect can better justify its proposal with the Client, presenting a rigorous environmental assessment of the project. Database source: ECOINVENT-KBOB	- Evaluates the energy balance, waste, emissions, resource depletion, acidification, eutrophication - Distinguish and separately evaluates the phases of construction, operation, renovation and demolition - Associated with Pleiade-Comfie	- Materials evaluated and not the products. - Lifespan limited to 90 years.	- The life of the building is limited to 90 years. Life components is accounted over that period, except for the interior finishes and exterior coatings that are 10 years old.
COCON	- Tools of comparison technical solutions in terms of environmental impact - Database and data expert INIES	- The tool is simple to use - Performance overview as radars	- The lifespan are typical. - No module economic evaluation and social	- Lifetimes typical products
BEES USA	- Méthode of Assessment of the environmental impact of products 	- Compare the impact of design choices both environmentally and economically - Algorithms and weights provided - Complete tool. - Only tool to provide information about the initial costs and future as well as a reflection on the update.	- No real study of the impact of lifespan.	The tool does not presents a true reflection on the impact of lifespan. Long lifespan is voluntarily limited to 50 years.

Appendix 3
Management Protocol Database on Environmental and Health Declaration of Construction Products (INIES)

A supervisory board is responsible for managing the database. Its mission is:
- Ensure the proper functioning of the database;
- To ensure continued compliance of the base vis- a- vis the development of the various regulations and standards relating to environmental and health messages;

The Board consists of representatives from various official institutions

Extract the admission procedure of EPD

The admission file a EPD INIES in the database must include:
- Complete EPD according to NF P01-010
- Evidence of the ability to use (conformity to standards, quality brands, technical approval ...
- The wording of the summary of the proposed EPD for health information and comfort

Regarding health information:
- Sanitary quality of interior spaces
- Domestic water quality
- Information comfort
- Certificate verification EPD issued in the framework of environmental and health declaration of construction products established by AFNOR, in the case of a EPD wishing appear as "checked" in the database INIES,
- The name and address of a responsible individual of EPD

M. Méquignon and H. Ait Haddou, *Lifetime Environmental Impact of Buildings*,
SpringerBriefs in Applied Sciences and Technology, DOI: 10.1007/978-3-319-06641-7,
© The Author(s) 2014